W0036545

SpringerBriefs in Biochemistry and Molecular Biology

More information about this series at http://www.springer.com/series/10196

Xin Liang • Landi Sun • Zhen Liu

Mechanosensory Transduction in Drosophila Melanogaster

 Springer

Xin Liang
Tsinghua-Peking Joint Center for Life
    Sciences, Max-Planck Partner Group,
    School of Life Sciences
Tsinghua University
Beijing, Beijing, China

Landi Sun
Tsinghua-Peking Joint Center for Life
    Sciences, Max-Planck Partner Group,
    School of Life Sciences
Tsinghua University
Beijing, Beijing, China

Zhen Liu
Tsinghua-Peking Joint Center for Life
    Sciences, Max-Planck Partner Group,
    School of Life Sciences
Tsinghua University
Beijing, Beijing, China

ISSN 2211-9353   ISSN 2211-9361   (electronic)
SpringerBriefs in Biochemistry and Molecular Biology
ISBN 978-981-10-6525-5   ISBN 978-981-10-6526-2   (eBook)
DOI 10.1007/978-981-10-6526-2

Library of Congress Control Number: 2017952942

Printed on acid-free paper

This Springer imprint is published by Springer Nature
The registered company is Springer Nature Singapore Pte Ltd.
The registered company address is: 152 Beach Road, #21-01/04 Gateway East, Singapore 189721, Singapore

Preface

The initial motivation to write this booklet was to prepare an introductory reading material for my students and postdocs who are going to work on mechanotransduction using fruit flies as their model organisms. Therefore, this booklet is designed to be elementary and hopefully reader-friendly. The organization of this booklet is very simple. I begin from the general roles of mechanotransduction in physiology (Chap. 1) and the basic considerations on the principles underlying the mechanotransduction process (Chap. 2). In this way, I hope to make connections for the readers from the physiological level to the microscopic level. In Chap. 3, I start to focus on fly mechanoreceptors by first introducing how the fly mechanosensitive cells contribute to the daily lives of flies, hoping to inspire the readers' curiosity on how these mechanoreceptors may work. Then, in Chap. 4, I provide the ultrastructural and mechanical details to explain the working mechanisms of various fly mechanoreceptors. Again, at the end of Chap. 4, my expectation is that the readers might be interested to further explore how the conversion of forces to intracellular signals may happen in the mechanosensitive cells, so I elaborate on the structural and functional details of the molecular transduction apparatus using NompC, a force-sensitive channel, as an example. By the end of this booklet, I hope the readers have had an overall idea about cell mechanotransduction and a handful of knowledge about fly mechanotransduction. Finally, I hope the readers enjoy reading this booklet and appreciate the structural-mechanical delicacy of fly mechanosensitive cells and molecules.

Acknowledgments

I would like to thank Fenghua Liu, who helped me in producing many of the figures in this booklet, and Jianfeng He, who assisted me in making the phylogenetic trees of NompC and Piezo. I would like to thank all the other members, Qixuan Wang, Wei Chen, Yinlong Song, and Hui Li, for their time in reading the earlier drafts and sharing their feedbacks. I would like to thank the School of Life Sciences (THU),

Tsinghua-Peking Joint Center for Life Sciences (THU), Max Planck Partner Group Program (THU & MPG), National Key R&D Program of China (2017YFA0503502) and National Natural Sciences Foundation of China (NSFC 31671389) for supporting my work. I also thank Dr. Peng Zhang (Springer, Beijing Office) for his support in the course of writing this booklet. Finally, I would like to thank my family, without whose support I would have never been able to write this booklet.

Beijing, China   Xin Liang

Contents

1 **Overview of Mechanosensory Transduction** 1
  1.1 Mechanosensory Transduction 1
    1.1.1 Sensory Transduction 1
    1.1.2 Mechanosensory Transduction in Physiology 2
    1.1.3 Dysfunction of Mechanotransduction in Diseases 2
    1.1.4 Summary 3
  1.2 Model I: Bacteria 4
    1.2.1 Mechanotransduction in Bacteria 4
    1.2.2 Mechanosensitive Channels (Msc) 5
    1.2.3 The Surface Attaching Apparatus 6
  1.3 Model II: Touch-Sensitive Neurons in *C. elegans* 6
    1.3.1 Mechanosensation in *C. elegans* 6
    1.3.2 A Good Animal Model 7
    1.3.3 Mechanosensory Neurons in *C. elegans* 8
    1.3.4 The Transduction Apparatus in *C. elegans* Touch
        Receptors 9
  1.4 Model III: Inner Ear Hair Cells 9
    1.4.1 Function and Structure of the Hair Cells 9
    1.4.2 Mechanotransduction Apparatus in Hair Cells 10
  1.5 Summary 11
  References 12

2 **"Gating-Spring" Model for Mechanotransduction** 13
  2.1 A Minimal Mechanotransduction Apparatus 13
    2.1.1 The Processer 13
    2.1.2 The Responder 15
    2.1.3 Examples 15
  2.2 The "Gating-Spring" Model 15
    2.2.1 Overview 15
    2.2.2 The Simple Mechanical Description 18

2.2.3   Model Predictions: Sensitivity and Dynamic Range   20
2.2.4   Functional Implications   24
2.2.5   Molecular Basis   24
References...   25

3   Mechanoreceptors in *Drosophila melanogaster*...............................   27
3.1   Overview of Fly Mechanoreceptors...   27
3.2   Type I Mechanoreceptors...   27
        3.2.1   Bristle Sensilla..   28
        3.2.2   Campaniform Sensilla   31
        3.2.3   Chordotonal Organ   33
        3.2.4   Ciliated Mechanoreceptors in Fly Larvae.............................   35
3.3   Type II Mechanoreceptors   36
        3.3.1   Class I da Neuron..   37
        3.3.2   Class II da Neuron   37
        3.3.3   Class III da Neuron   38
        3.3.4   Class IV da Neuron...   39
3.4   Summary...   40
References...   40

4   Mechanotransduction in *Drosophila* Mechanoreceptors.....................   43
4.1   Overview of Fly Mechanotransduction   43
4.2   Bristle Receptor   44
        4.2.1   Bristle Deflection   44
        4.2.2   Dendritic Tip and the Supporting Structures   47
4.3   Campaniform Receptor..   48
        4.3.1   Cuticle Deformation   48
        4.3.2   Dendritic Tip and the Supporting Structures   50
        4.3.3   Molecular Basis of Mechanotransduction   52
4.4   Chordotonal Organ   54
        4.4.1   Fly Antenna and Johnston's Organ   55
        4.4.2   Molecular Basis of Mechanotransduction   55
4.5   Dendritic Arborization Neurons   57
        4.5.1   Overall Mechanics of da Neurons   57
        4.5.2   Molecular Basis of Mechanotransduction   58
References...   59

5   *Drosophila* Mechanotransduction Channels......................................   63
5.1   Overview..   63
5.2   No Mechanoreceptor Potential C (NompC)   64
        5.2.1   Overview on Fly NompC..   64
        5.2.2   Structure of NompC   65
        5.2.3   Gating of NompC   67
        5.2.4   The Gating Spring of NompC..   68
        5.2.5   NompC-Microtubule Interaction   71

    5.2.6   Physiological Roles of Fly NompC 72
    5.2.7   NompC/TRPN in Other Organisms....................................... 73
  5.3   DmPiezo 74
  Appendix.. 76
  References.. 78

Afterword.. 81

Chapter 1
Overview of Mechanosensory Transduction

Abstract As the start of this booklet, we begin this chapter by providing a definition on "what is mechanosensory transduction" at the cell biological level and listing the fundamental questions in this emerging field. We then briefly introduce "mechanotransduction in physiology" and "mechanotransduction in diseases," which aim to integrate this cell biological process into the physiological and pathological contexts. Finally, we summarize the mechanotransduction processes in three model systems: (1) the bacterial cell, (2) the *C. elegans* touch-sensitive receptor, and (3) the vertebrate hair cell. We focus on how mechanotransduction contributes to the functions of these cells and the molecular basis of the mechanosensory transduction in each model cell.

1.1 Mechanosensory Transduction

1.1.1 Sensory Transduction

Organisms interact with the external environment by sensing a wide range of signals, including chemical substances, electromagnetic waves, mechanical forces, etc. These environmental signals need to be converted or transduced to the signals that can be read by the organisms (Chalfie 2009) or, more specifically, by the sensory cells. The nature of the signal transduction process is key to understand the theoretical and operational mechanisms of the sensory systems. In Box 1.1, the fundamental questions in the studies of sensory transduction are listed, and they apply well to mechanotransduction. Interestingly, although all sensory processes are important to our life, our understandings of different sensory transductions are at vastly different levels. For example, rhodopsin, a light transducer in phototransduction, was already discovered in nineteenth century (by Franz Christian Boll in 1876), while the protein that transduces sound in the hearing system still remains mysterious to date (Zhao and Muller 2015).

© The Author(s) 2017
X. Liang et al., *Mechanosensory Transduction in Drosophila Melanogaster*,
SpringerBriefs in Biochemistry and Molecular Biology,
DOI 10.1007/978-981-10-6526-2_1

Fig. 1.1 Mechanosensory transduction is the process that transforms the mechanical signals in the environment into cellular signals

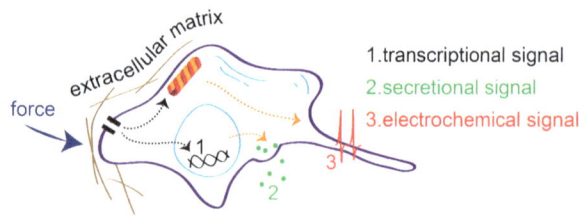

Box 1.1: The Fundamental Questions in Sensory Transduction
1. What are the physical and chemical principles that govern the signal transduction?
2. What are the molecular transducers?
3. How are the molecules organized spatiotemporally to transduce the signals?
4. How are the transduction processes regulated in cells?

1.1.2   Mechanosensory Transduction in Physiology

Among various external signals, the mechanical stimuli in the surroundings constantly arrive at the membrane of a cell. The process that a cell converts these mechanical stimuli into electrochemical, secretional, or transcriptional signals is termed as **mechanosensory transduction** (mechanotransduction) (Gillespie and Walker 2001) (Fig. 1.1). All organisms/cells physically interact with their surroundings by sensing various mechanical signals with different spatiotemporal patterns and intensities. In Box 1.2, three examples are listed to demonstrate the important roles of mechanotransduction in the organisms of different complexities. These examples show that mechanosensory transduction is a fundamental process that matters in general physiology. However, our understanding on mechanotransduction is the least compared to the other sensory processes, such as vision, smell, and taste (Chalfie 2009).

1.1.3   Dysfunction of Mechanotransduction in Diseases

Like all the other physiological processes, the dysfunction of mechanotransduction has been found in several human diseases (Jaalouk and Lammerding 2009), including deafness, arteriosclerosis, polycystic kidney disease, muscular dystrophies, etc. In Box 1.3, we provide two examples of the mechanotransduction-related diseases. We first introduce the function of a specific mechanosensory system in normal physiology and then describe its dysfunction in the relevant diseases. More information about the mechanotransduction and diseases can be found elsewhere (Jaalouk and Lammerding 2009).

Box 1.2: Mechanosensory Transduction in Different Organisms

Unicellular organisms, like *E. coli*, are able to sense the changes in the osmolarity of the surrounding fluid to facilitate the osmolarity adaptation (Kung et al. 2010).

Invertebrates, like flies and worms, can sense cuticle deformation while crawling, walking, or flying, which provides feedback signals for their motility control (Gillespie and Walker 2001; Goodman and Schwarz 2003; Goodman 2006).

Mammals, like us, can hear others through the perception of sound, feel others through the perception of touch, and balance our bodies though the perception of acceleration (Gillespie and Walker 2001; Chalfie 2009; Gillespie and Muller 2009; Lumpkin et al. 2010).

Box 1.3: Hearing Loss and Muscular Dystrophies

Hearing loss: The changes in sound pressure are normally detected by the hair cells in the inner ear. The key mechanotransduction micromachinery of hair cells locates at the mechanoreceptive organelle of the hair cells, the hair buddle. The mutations in the mechanosensitive molecules lead to the dysfunction of the mechanotransduction apparatus and thereby contribute to human deafness (Vollrath et al. 2007).

Muscular dystrophies: Not only in the specialized sensory cells (e.g., hair cell), mechanotransduction is also important for a broad range of tissues and cells. For example, forces are generated by the sarcomere components in skeletal muscle cells and transmitted between the muscle cell and the extracellular matrix through a protein complex that contains dystrophin and the dystrophin-associated proteins. This mechanical design ensures the generation of forces that fit the load to protect the sarcomere membrane from being ruptured due to the excessive forces. However, the mutations in dystrophin genes, sarcomere component genes, and several nuclear proteins lead to the abnormality in the sarcomere mechanics, reduction in the force generation or transmission, and finally the degeneration or death of the muscle cells (Jaalouk and Lammerding 2009).

1.1.4 Summary

How do cells sense mechanical stimuli? Although we mainly discuss the mechanotransduction in *Drosophila* models in this booklet, this question will stand as the center throughout the book. In the first chapter, we begin by shortly introducing the classical model organisms in the studies of mechanotransduction. These models range from unicellular organisms to more complicated organisms, e.g., vertebrates.

1.2   Model I: Bacteria

1.2.1   Mechanotransduction in Bacteria

Bacteria have a simple form of life, but their long evolutionary history endows them with the ability to respond to a wide range of environmental stimuli, including chemical, thermal, electromagnetic, and mechanical signals. They evolve the sensory systems that can quickly adapt themselves to the changes in their nature habitats. The remarkable plasticity of bacterial sensory systems contributes to several amazing facts, among others, of bacteria: (1) a large number of bacterial cells on Earth. For example, there are 40 million bacterial cells per gram of soil, and the biomass of bacteria on earth overtakes the sum of all plants and animals. (2) Bacteria develop an extremely broad diversity in the long evolutionary path. They inhabit in our daily environment but are also found in extreme conditions, like acidic hot springs, radioactive wastes, the deepest parts of the oceans, the deep crust, and the polar regions of the Earth.

Just like all the other signals, mechanical stimuli are present in the daily life of bacteria in soil and water. However, the mechanical world around a bacterial cell and its influence on the behavior of bacteria only start to be recognized recently (Persat et al. 2015). In Box 1.4, we present two little stories about a bacterial cell and its surrounding mechanics. In the following Sects. 1.2.2 and 1.2.3, we elaborate on the key molecules that help bacteria deal with the mechanical stresses.

Box 1.4: Two Examples of Mechanics Around a Bacterial Cell

Osmotic shock: In a rainy day, a downpour washes a bacterial cell out of the soil. While this cell is swimming in the rainwater, it is actually experiencing a hypotonic shock. The rapid dilution of the environmental osmolytes leads to an osmotic force that creates an increase in the membrane tension, which could potentially break the cell membrane and end the life of this cell. In order to survive, this cell opens the mechanosensitive channels (Msc) which sense the increase in the membrane tension and respond by releasing cytoplasmic osmolytes, thereby relieving the imbalanced osmotic pressure and protecting the cell membrane from being ruptured (Kung et al. 2010).

Fluid flow: In a vast ocean, a bacterial cell attaches to a surface (of an animal or plant tissue). However, the nonstop flow in ocean constantly generates the shear stress at the solid-liquid interface that is strong enough to flush this cell away from the surface. This cell likes this habitat, so it secretes adhesive substances to construct its holdfasts which produce enough drag forces that maintain the attachment to the substratum. More safely, the mechanosensory system of this cell is able to measure the shear stress and adjust its adhesive forces accordingly. In such a way, this cell could remain attached on a solid surface in a variety of flow environments (Persat et al. 2015; Persat 2017).

1.2.2   Mechanosensitive Channels (Msc) (Kung et al. 2010)

To instantly release the cytoplasmic osmolytes upon the sudden hypotonic shocks, *E. coli* cells recruit a set of protein channels as their safety valves. These channels are both mechanosensors and mechanoresponders, namely, they sense the increase in the membrane tension and respond by opening their channel pores which are gateways for the osmolytes. They are generally termed as mechanosensitive channels (Msc) (Fig. 1.2a). In *E. coli*, a set of Mscs constitutes an osmolyte-release system. These channels include (1) MscL, an Msc with the **largest** conductance (3 nS) in all four Mscs; (2) MscM, an Msc with the **minimal** conductance (0.3–0.4 nS); (3) MscK, a **K**$^+$-dependent Msc with an intermediate conductance (1 nS); and (4) MscS, an Msc with a relatively **small**/intermediate conductance (1 nS). These channels provide step responses based on the levels of the osmotic stresses. MscM has the lowest threshold, so it acts as the first responder, but its small conductance only allows the release of small amount of osmolytes. MscK and MscS, with the intermediate thresholds, respond to the increased level of osmolarity-down stress, with the intermediate conductance. MscL, with the highest threshold (nearly the lytic limit) but largest conductance, functions as the ultimate safety valve for *E. coli* cells by releasing a large amount of osmolytes through its huge channel pore (3 nm in diameter, 1.5 nm change upon opening, largest known conformational change in channels). Given the expensive metabolic costs associated with the gating of nonselective Mscs, this stress-dependent osmolyte-release system ensures a green (energy saving) mechanotransduction system that rapidly responds to the osmotic perturbations in the environment.

Fig. 1.2 Two examples of mechanotransduction in bacteria. (**a**) Msc channels in the relief of the hypotonic shock Kung et al. (2010). The colors of the channels encode for their different thresholds and conductance. (**b**) A bacterial cell attaches to a surface using the fimbriae Persat et al. (2015), Persat (2017). The FimH-mannose complex is illustrated and enlarged in the *dashed circle*

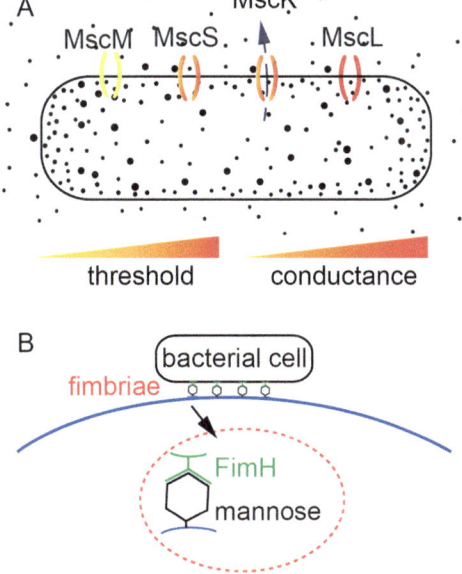

1.2.3 The Surface Attaching Apparatus (Persat et al. 2015; Persat 2017)

To resist the shear stress near the solid-liquid interface and remain attached to the surface, *E. coli* cells develop an adhesive linkage apparatus, the fimbriae (Fig. 1.2b). A typical fimbria has a diameter of a few nanometers and can be as long as a few microns. They carry adhesins which attach the bacterial cell to the other cell or the surface of various tissues or some inanimate objects. A well characterized adhesin protein is FimH that binds to the D-mannose present on the surface of many tissues. A remarkable feature of the FimH-mannose connection is that when the shear stress exerts an increasing tensile load onto the fimbriae (mechanosensitive), the mannose-bound FimH undergoes a conformational change that supports a stronger linkage (mechanoresponsive), thereby improving the stability of the cell attachment to the surface. This mechanism provides a force-dependent attachment that is reliable and flexible. Such a flow-dependent attaching mechanism is not rare among various bacterial strains and may be achieved by recruiting different adhesins.

1.3 Model II: Touch-Sensitive Neurons in *C. elegans*

1.3.1 Mechanosensation in C. elegans

C. elegans is a small (about 1 mm long, only a thousand times longer than a bacterial cell) nematode. It has a simple body organization of only 959 cells in which 302 are neurons. Wild *C. elegans* lives in soil or dirt and it eats bacteria. In the soil environment (three dimensional, damp, dark, full of obstacles), *C. elegans* crawls in the soil. Just like the bacterial cell, it experiences a mechanical world, for example, when colliding with small particles or running into the other animals. Therefore, several important behaviors (e.g., feeding, mating, and locomotion) are initiated or dependent on the mechanosensation processes. Despite the simplicity, *C. elegans* develops a set of mechanosensory cells that respond to a range of mechanical signals in its nature habitat. The simple body organization, the transparency of the cuticle, the powerful genetic manipulation tools, and the direct manifestation of mechanosensitivity through the behavior, among other reasons, make *C. elegans* a great model to study the general principle and molecular basis underlying the mechanosensation (Kamkin and Kiseleva 2005).

Box 1.5: Behavior Assays in the Studies of *C. elegans* Mechanosensitivity (Hart 2006)

Nose touch: As a worm moves forward, it makes contact with an obstacle (e.g., a hair placed in the testing plate) using its nose tip. Upon contact, normal animals immediately initiate backward movement, but the defective animals might crawl over the obstacle or slide along the obstacle. Each individual worm should be tested repeatedly for ten times to get a score of nose touch response for this worm.

Gentle touch: *C. elegans* senses a variety of gentle touches, including the gentle touches to the body, head, tail, and nose, so there are a set of behavior assays for these touches. One qualitative assay is to mount an eyebrow hair to a toothpick and stroke the animal body using this hair. The touch-sensitive animals respond by stopping the movement or moving away from the hair, while the touch-insensitive animals fail to respond to such a stimulus. This assay can be quantitative by scoring the responses to multiple touches. The other more qualitative assay is to examine the response of worms to "tapping" stimuli. In this assay, one taps the plate and such taps create vibrations in worm's environments. The touch-sensitive animals retreat for some distance, while the touch-insensitive animals fail to respond to this stimulus. This is a rapid assay without a high accuracy compared to the hair-based method, but it was successfully applied in screening for touch-insensitive mutants.

Harsh touch: The worms that are insensitive to gentle touches might respond to the stronger forces, i.e., "harsh touches." In this assay, the worm's response is measured by poking or prodding the animal with a platinum wire. The normal animals respond by initiating locomotion, while the defective animals (or starving animals) fail to give any visible responses.

1.3.2   A Good Animal Model

Because many behaviors of *C. elegans* directly reflect its mechanosensitivity, a set of behavior assays were established to examine the mechanosensitivity of *C. elegans* to various force stimuli, and we introduce some often-used assays in Box 1.5.

Convenient genetic manipulations add up to the success of *C. elegans* in studying mechanosensitivity. This is originally powered by a set of classic genetic methods, including random mutagenesis, precise genetic mapping, and simple gene cloning (Kamkin and Kiseleva 2005). In the last few years, this has become technically even more straightforward, thanks to the development of the Crispr/cas9 systems (Li and Ou 2016).

1.3.3 Mechanosensory Neurons in C. elegans

In adult hermaphrodites, mechanosensory cells (30 cells) are ciliated or non-ciliated. Adult males have 52 additional ciliated sensory neurons that are important in the mating behaviors. The ciliated neurons have their dendrite encapsulated within the cuticle or exposed to the external environment, while the non-ciliated neurons are inside the animal's body. These sensory cells are responsive to various mechanical stimuli mentioned in Box 1.5. A detailed summary can be found elsewhere (Goodman and Schwarz 2003; Kamkin and Kiseleva 2005; Goodman 2006). Here, we take the touch receptors as an introductory example because these cells are among the best-studied mechanosensory cells in the C. elegans mechanosensory system.

There are six touch receptors (ALML, ALMR, AVM, PLML, PLMR, and PVM, see Box 1.6) in C. elegans (Fig. 1.3a). They share a common plan for their neuronal morphology; each of these six neurons has a long dendrite that extends nearly one-half length of the worm's body; these dendrites are filled with large 15-protofilament microtubules that are unusual and unique to the touch receptor neurons. Among these six cells, ALML, ALMR, and AVM respond to the touches onto the anterior part of the worm; PLML and PLMR are responsible for the stimuli onto the posterior part; the exact roles of PVM have not been demonstrated yet so far. These cells are sensors for the eyebrow hair strokes and the plate "tapping" signals, namely, the gentle touches to the worm's body wall. These touch cells have been studied in depth with the genetic and electrophysiological methods. We will summarize the molecular basis of the mechanotransduction in the touch cells shortly.

Fig. 1.3 Mechano-transduction apparatus in the touch receptors of C. elegans. (**a**) Touch receptors in C. elegans. Only the *left* side of the worm is illustrated, and only the dendrite of ALML is shown. The nose touch and body touch stimulation are shown with two *arrows* (The pictures are drawn and modified after the reference Goodman 2006). (**b**) The simplified cartoon for the mechanotransduction apparatus in the touch receptors (Modified after the reference Lumpkin et al. 2010)

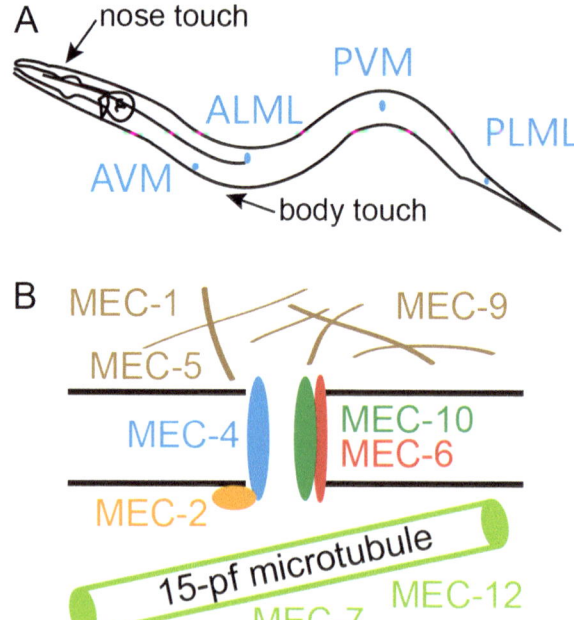

Box 1.6: Abbreviations of the Touch Receptors
ALML/ALMR: Anterior lateral microtubule cell left/right
AVM: Anterior ventral microtubule cell
PLML/PLMR: Anterior lateral microtubule cell left/right
PVM: Posterior ventral microtubule cell

1.3.4 The Transduction Apparatus in C. elegans Touch Receptors (Goodman 2006; Lumpkin et al. 2010)

Early genetic screens for the touch-insensitive mutants identified a set of 12 *mec* genes involved in the mechanosensory functions of the *C. elegans* touch receptors, from which the MEC-4 complex was demonstrated to be a mechanotransduction channel and is to date one of the best characterized eukaryotic mechanotransduction channels (Fig. 1.3b). The core of this transduction channel complex is formed of MEC-4 and MEC-10, both of which are DEG/ENaC subunits. Two regulators, MEC-2 and MEC-6, largely enhance the channel activities of MEC-4/MEC-10 complex. These four proteins are essential components of the transduction channel complex, and mutations in any of these proteins lead to the failure in touch-evoked behaviors. Apart from the core components, the channel complex is structurally supported by the extracellular matrix outside the cells (MEC-1, MEC-5, and MEC-9), as well as the cytoskeletal structures inside the cells (MEC-7 and MEC-12). Mutations in the matrix proteins and the cytoskeletal proteins lead to the disruption of mechanotransduction to different extents.

1.4 Model III: Inner Ear Hair Cells

1.4.1 Function and Structure of the Hair Cells (Gillespie and Muller 2009)

Hair cells locate in the auditory and vestibular systems of the vertebrate inner ears. They are mechanosensory cells that are able to convert mechanical stimuli into electrical responses. This transduction process is key in our perceptions of sound, head movement, and fluid motion; therefore it is important in human physiology. The hair cell has a specialized mechanoreceptive organelle, the hair bundle. It resides on the apical surface of the hair cell and is a compact cluster of stereocilia that are filled with the F-actin fibers. The deflection of the hair bundle, caused by the fluid motion or structural movement in the inner ear, excites the hair cells and thus is the first step in generating the auditory responses (Fig. 1.4a). Hair bundle is extremely sensitive to the mechanical stimuli (e.g., ~1 degree angular deflection). This high sensitivity

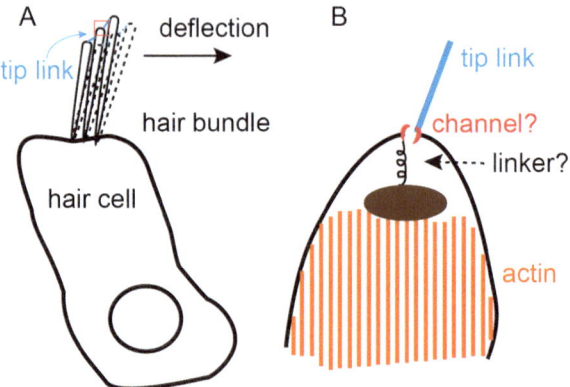

Fig. 1.4 Hair cell mechanotransduction. (**a**) The cartoon for a hair cell in which the deflection of the hair bundle is illustrated with an *arrow*. The part highlighted in the *red box* is enlarged in (**b**). (**b**) The enlarged cartoon for the transduction apparatus at the base of a tip link. The possible components in this molecular complex are shown as indicated

is, at least partially, attributed to the bundle's amazing architecture that provides the structural-mechanical basis for the mechanotransduction. Hair bundle consists of ~100 stereocilia (varies in different species). The tips of these stereocilia are connected to their taller neighbors via a fine extracellular filamentous structure, the tip link. The tip links align the stereocilia to the mechanosensitive axis of the hair bundle as an entirety and position them appropriately to participate in the mechanotransduction when the hair bundle is deflected.

1.4.2   Mechanotransduction Apparatus in Hair Cells

The hair bundle recruits a set of incompletely identified/characterized proteins to transduce the mechanical signals into an electrical one (Gillespie and Muller 2009; Zhao and Muller 2015; Hudspeth 2014). These molecules form a transduction apparatus that resides in the plasma membrane of the shorter stereocilium (in the pair linked by a tip link) at the base of the tip link (Fig. 1.4B). Structurally, this transduction apparatus consists of the base of a tip link, the plasma membrane, the putative transduction channel complex in the membrane, possibly an intracellular linker, the actin-based cytoskeleton structure, and the accessory cytoskeleton-based structures (e.g., motors). This complex forms a structural linkage that provides a mechanical pathway to transmit the tension on the tip links, caused by the relative movement of the adjacent stereocilia in the tip-link-linking pair due to the hair bundle deflection, to the transduction channels in the plasma membrane.

Mechanically, the hair bundle is a strain gauge that measures the deflection and responds to it, but the mechanics that governs the operation of this molecular gauge is unclear (Hudspeth 2014). It was proposed that there're spring-like structures that provide gating-associated mechanical compliance in this transduction system based on the mechanical and electrophysiological experiments. However, the structural components that determine the internal mechanics of this molecular apparatus remain unknown. This structural-mechanical determinant sets the force loaded onto the transduction channel and thereby controls the gating behaviors of the channels (see Chap. 2). The candidates for such a spring-like component include the filamentous tip links, the plasma membrane, and the intracellular linkers (if they exist), but the ultimate clarification on this issue depends on the discovery of the molecular identities of these structures, the precise measurements of their stiffness, and the establishment of a structural-mechanical model of the hair bundle (Powers et al. 2012; Zhao and Muller 2015).

The key step in understanding the transduction mechanism in hair cells is to identify the molecular identities of the components in the transduction apparatus. Despite the efforts made in the last 30 years, the complete reveal of the transduction molecules is still underway. Nevertheless, much progresses have been made in the last years. For example, the tip links have been found to be made of CDH23 and PCDH15; four integral membrane proteins, LHFPL5/TMHS, TMIE, TMC1, and TMC2, are found to be closely related to the transduction channels though none of them was found to be a pore-forming subunit; Myo15-SANS and Myo7a-SANS were shown to concentrate on the upper and lower tip link ends, respectively. How do these molecules contribute to the transduction remains elusive, and the identification of the transduction channel, in particular, still resides in the center of the mystery of the hair cell mechanotransduction (Gillespie and Muller 2009; Zhao and Muller 2015).

1.5 Summary

In the last Sects. 1.2, 1.3 and 1.4, three model systems that were often used in the studies of mechanotransduction were introduced. The rest of this booklet will focus on another, yet important, model system, the fly mechanoreceptors. In the next chapters, we will elaborate on the cell biological, molecular, structural, and mechanical basis of fly mechanotransduction. Nevertheless, I think that the most important thing, whichever model system in use, is to learn the physical principles that govern the operation of a mechanosensory system and then how these principles are biologically engineered in a specific living organism/a model cell. Based on these thoughts, we start the discussion on fly mechanotransduction by first introducing the general model for mechanotransduction in Chap. 2.

References

Chalfie M (2009) Neurosensory mechanotransduction. Nat Rev Mol Cell Biol 10(1):44–52. https://doi.org/10.1038/nrm2595

Gillespie PG, Muller U (2009) Mechanotransduction by hair cells: models, molecules, and mechanisms. Cell 139(1):33–44. https://doi.org/10.1016/j.cell.2009.09.010

Gillespie PG, Walker RG (2001) Molecular basis of mechanosensory transduction. Nature 413(6852):194–202. https://doi.org/10.1038/35093011

Goodman MB (2006) Mechanosensation. WormBook:1–14. https://doi.org/10.1895/wormbook.1.62.1

Goodman MB, Schwarz EM (2003) Transducing touch in Caenorhabditis elegans. Annu Rev Physiol 65:429–452. https://doi.org/10.1146/annurev.physiol.65.092101.142659

Hart A (2006) Behavior. WormBook. https://doi.org/10.1895/wormbook.1.87.1

Hudspeth AJ (2014) Integrating the active process of hair cells with cochlear function. Nat Rev Neurosci 15(9):600–614. https://doi.org/10.1038/nrn3786

Jaalouk DE, Lammerding J (2009) Mechanotransduction gone awry. Nat Rev Mol Cell Biol 10(1):63–73. https://doi.org/10.1038/nrm2597

Kamkin AE, Kiseleva IE (2005) In: Kamkin A, Kiseleva I (eds) Mechanosensitivity in cells and tissues. Academia, Moscow

Kung C, Martinac B, Sukharev S (2010) Mechanosensitive channels in microbes. Annu Rev Microbiol 64:313–329. https://doi.org/10.1146/annurev.micro.112408.134106

Li W, Ou G (2016) The application of somatic CRISPR-Cas9 to conditional genome editing in Caenorhabditis elegans. Genesis 54(4):170–181. https://doi.org/10.1002/dvg.22932

Lumpkin EA, Marshall KL, Nelson AM (2010) Review series: the cell biology of touch. J Cell Biol 191(2):237–248. https://doi.org/10.1083/jcb.201006074

Persat A (2017) Bacterial mechanotransduction. Curr Opin Microbiol 36:1–6. https://doi.org/10.1016/j.mib.2016.12.002

Persat A, Nadell CD, Kim MK, Ingremeau F, Siryaporn A, Drescher K et al (2015) The mechanical world of bacteria. Cell 161(5):988–997. https://doi.org/10.1016/j.cell.2015.05.005

Powers RJ, Roy S, Atilgan E, Brownell WE, Sun SX, Gillespie PG et al (2012) Stereocilia membrane deformation: implications for the gating spring and mechanotransduction channel. Biophys J 102(2):201–210. https://doi.org/10.1016/j.bpj.2011.12.022

Vollrath MA, Kwan KY, Corey DP (2007) The micromachinery of mechanotransduction in hair cells. Annu Rev Neurosci 30:339–365. https://doi.org/10.1146/annurev.neuro.29.051605.112917

Zhao B, Muller U (2015) The elusive mechanotransduction machinery of hair cells. Curr Opin Neurobiol 34:172–179. https://doi.org/10.1016/j.conb.2015.08.006

Chapter 2
"Gating-Spring" Model for Mechanotransduction

Abstract In Chap. 1, we provided a general introduction for mechanotransduction. We feel that before going to the biological details (molecule, structures, etc.), it may help to provide the readers an overall picture on how an ideal system may work. We think that a theoretical model that depicts the operation of a mechanotransduction system may serve this purpose well. In this chapter, we introduce a classical model for cell mechanotransduction, i.e., the "gating-spring" model, which was proposed over 30 years ago based on the remarkable experimental work and data analysis on hair cells. We add some of our understanding of this model at the end of this chapter.

2.1 A Minimal Mechanotransduction Apparatus

Intuitively, the transduction apparatus in the cell should contain at least two components: a processer and a responder (Fig. 2.1). The processer is a linkage through which the distal stimuli (environmental signals) are processed and then transmitted to the responder, basically acting as a mechanical signaling pathway. The responder is a transducer that converts the proximal mechanical stimuli into a cellular signal, e.g., the electrical current. Based on this conceptual design, different cells may develop different configurations that include other supportive and regulatory elements for different physiological functions.

2.1.1 The Processer

In tissues, the cellular responders do not usually have a direct contact with the environmental signals (i.e., the distal stimuli like sound and touch). In fact, the distal stimuli are transmitted to the responder molecules via other structures (membrane, intracellular cytoskeletons, and extracellular matrix). While being transmitted, the signals are also processed. The signal processing varies in different cells. A possible scenario is that the processer could be thought of as a **spring** that converts a deformation signal into a force (Fig. 2.1b). If the spring has a constant stiffness, the distal deformation is linearly transformed into a proximal force ($f = \kappa \cdot D$). That means for

© The Author(s) 2017

X. Liang et al., *Mechanosensory Transduction in Drosophila Melanogaster*, SpringerBriefs in Biochemistry and Molecular Biology, DOI 10.1007/978-981-10-6526-2_2

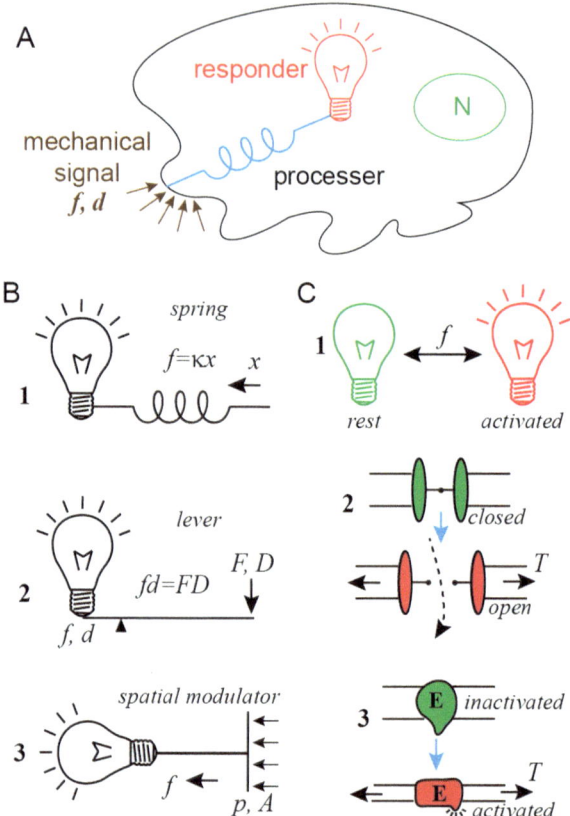

Fig. 2.1 Simplified schematics of the mechanotransduction apparatuses. (**a**) A conceptual minimal transduction apparatus consists of a processer and a responder in cells. (**b**) Three examples of the processers. (*1*) A spring-like processer that converts the deformation (x) into a force (f). (*2*) A lever-like processer that transforms the large deformations (D)/small forces (F) into small deformations (d)/large forces (f). (*3*) A spatial modulator that focuses the pressure (p) onto a surface (A) into a compressive force to the responder (f). (**c**) The responders. (*1*) The responder's transition from one state to another occurs under a force load. (*2*) Channels open in response to an increase in the membrane tension (T). (*3*) Enzymes are activated when membrane tension increases (T)

a given distal stimulus (D), a stiffer spring will generate a larger force. Another scenario is that the processer acts as a **lever** that transforms the bigger distal deformation into a smaller proximal change and in the meantime amplifies the smaller distal force into a bigger proximal force (Fig. 2.1b). It is also possible that the processer acts like a spring combined with a lever that transforms and amplifies the signals. In addition, if the stiffness of the springlike processer is dependent on the stretched or compressed state of the spring-like components, the distal stimuli are then nonlinearly transformed into the proximal ones. The processer could also act as a **spatial modulator** that modifies the distribution of the distal stimuli (Fig. 2.1b). One example is the tip links in hair cells that focus the deflection of the entire hair bundle (see Chap. 1, Fig. 2.4) onto the tips of the stereocilia. Another example is the microtubule-membrane connectors in fly mechanoreceptors, which we will discuss in details later (see Chap. 4). The processer might also function as a **frequency analyzer** which reads the temporal characteristics of the mechanical stimuli. The possible examples might exist in vertebrate hair cells and the fly hearing organ (i.e., the Johnston's organ), which however still remain further elusive.

2.1.2   The Responder

The responder is a transducer that converts the proximal but yet mechanical signals finally into cellular signals (e.g. changes in the electrical current or enzymatic activity) (Fig. 2.1c). Ion channels could be the responder (Fig. 2.1c). Channels respond to the forces by changing their conformations (e.g., force-dependent open/close of the channel pores), and such structural changes, in turn, lead to the changes in the electrical current through the channel pores (Markin and Hudspeth 1995). This process converts the forces or deformations into an electrical signal. This process is most often seen in the specialized sensory neurons where the changes in the current lead to the depolarization of the neuronal membrane and then trigger the neuronal impulses. Enzymes could be the responders as well (Fig. 2.1c). Just like the ion channels, the enzymes respond by changing the enzymatic activities (e.g., GPCRs), which then biochemically tunes the activities of the intracellular signaling pathways and thereby converts the mechanical signals into the chemical signals in the cells.

2.1.3   Examples

The processer and the responder together form the core of the transduction apparatus. The processer transmits the signal and matches the differences in the mechanical impedance between the distal signal receiver (e.g., hair bundle in hair cells) and the responder (e.g., mechanotransduction channels). The responder ensures that the cells are able to make sensible responses to the external stimuli. Because each type of the mechanosensory cells is specialized for a certain class of mechanical stimuli (based on the intensity, spatial distribution, frequency, etc.), it is intuitive to hypothesize that each type of sensory cell recruits a specific pair of a processer and a responder (consider the three model examples in Chap. 1, what are the processors and responders?).

2.2   The "Gating-Spring" Model

2.2.1   Overview

("What is the direct gating model?") The "gating-spring" model depicts that mechanotransduction occurs as the mechanical force is transmitted through an elastic element to the transduction channel. The changes in the forces in the elastic element switch the channel between the open and the closed states. In fact, this model is a specific extension of a more general model which posits that the mechanotransduction channels are gated directly by forces (i.e., **the direct gating model**). The direct gating model was initially proposed based on the experimental observation of the very short latency (~ 40 µs) between mechanical stimuli and the electrophysiological responses in the bullfrog hair cells (Corey and Hudspeth 1979) (and later in other mechanosensory cells (Albert et al. 2007; Liang et al. 2014; O'Hagan et al. 2005; Walker et al. 2000). Within this short period of time, a small signal molecule is only able to diffuse in the range of a few nanometers (the size of a not-so-big protein molecule) in cytosol (Howard 2001), so random diffusion is too slow to account for the rapid mechanical signaling. Based on this idea, it was thought that there is most likely a direct gating mechanism, but not the conventional diffusion-based signaling mechanism, that underlies the mechanotransduction.

("What is the gating-spring model in particular?") In the direct gating model, the mechanical signaling pathway is a set of cellular structures that transmits the distal mechanical signals to the responder molecules. **The "gating-spring" model** extends the direct gating mechanism by proposing that the mechanical signaling pathway has a stiffness at the order of a few pN/nm. This stiffness suggests that the mechanical signaling pathway is relatively compliant compared to the rigid structures in cells, for example, microtubules and collagens, and at the order of the soft structures in cells, for example, cell membrane and molecular motors (e.g., myosin). Therefore, the "gating-spring" model suggests that the mechanical signaling pathway does not only transmit the force but also provides the mechanical compliance to the system, mechanically as a spring-like processer. In fact, this compliance is inherently required in the direct gating mechanism. This is because the mechanical stimuli in the environment usually cause a relatively large deformation (a few microns or at least hundreds of nanometers) of the tissues (e.g., when a finger touches the skin, the pit is visible, Fig. 2.2). However, the conformational changes of the responders (i.e., channels or enzymes) in cells are only in the order of a few nanometers or even smaller. Therefore, there must be soft components in the mechanical signaling pathway that take up much of the deformation and, by doing so, regulate the conformational changes of the responder molecules. In this way, the mechanical compliance in the transduction apparatus ensures the force-dependent gating of the responder, namely, that the response of the channel is graded rather than bimodal (i.e., all-or-none).

("What is 'gating spring'?") The source (i.e., the material/structures/components in the mechanical signaling pathway) of the mechanical compliance is iconically termed as the "**gating spring.**" Formally, the "gating spring" represents the lumped stiffness of all components in the mechanical signaling pathway because every material contributes some compliance. In practice, it is the relative compliance that matters. For example, if one component is significantly more compliant than the others, this component determines the overall compliance of the signaling pathway. In this scenario, "gating spring" is also taken to mean the most compliant component(s).

Fig. 2.2 The touch on the skin activates the sensory molecules in the touch-sensitive cells (Note that the deformation of the skin is much larger than the conformational change of the sensory molecules (an ion channel in this case). The soft materials (i.e., the *upper* spring) between the skin tissue and the ion channel must have taken up much of the deformation when transmitting the forces)

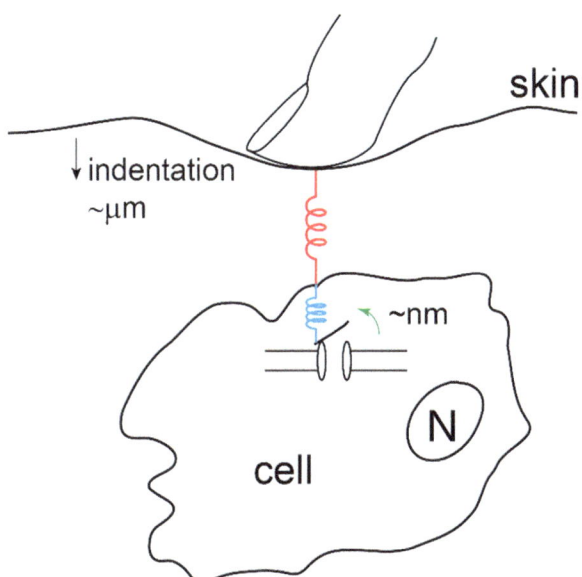

("Why is the 'gating spring' important?") The molecular identification of the gating springs is important in understanding the mechanical and molecular basis of mechanotransduction. We could consider this issue by comparing the mechanical signaling process to the conventional biochemistry-based signaling pathways (Fig. 2.3). In a simple example of the biochemistry-based signaling process, a series reactions occur in the step-by-step manner from the first binding of a ligand to its receptor until the final changes in gene transcriptions. The key reaction that determines the rate of the entire signaling process is usually called the "rate-limiting" step and has the slowest reaction rate. Mechanical signaling pathways are not based on biochemical reactions but physical interactions (according to the direct gating model). Despite the differences in the signaling forms, both can be considered as the linear signaling processes. In the mechanical schematic, the gating springs are the most compliant structures in the mechanical signaling pathways. Just like the "rate-limiting" reaction that determines the overall rate of signaling in the biochemistry-based signaling pathways, the gating spring determines the overall stiffness of the mechanical signaling pathway and finally determines the forces applied onto the responder molecules (see Eq. 2.6 below). In this view, the gating spring is the core of the mechanical signaling pathway, while the other rigid structures can be seen as the supportive or linkage components. As we will see in the next sections, the compliance of the "gating spring" quantitatively shapes the signaling performance of the mechanosensory system in response to the stimuli.

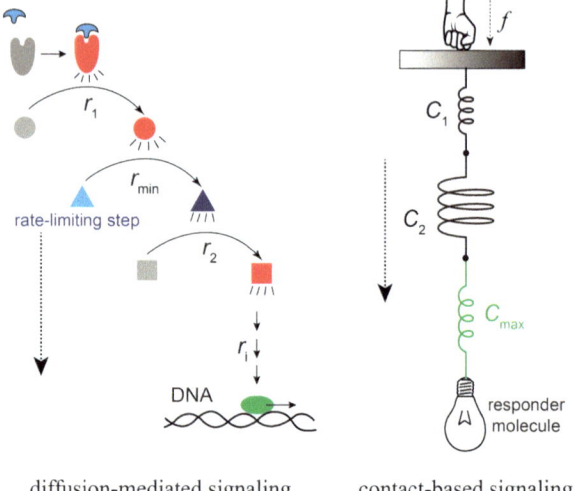

diffusion-mediated signaling   contact-based signaling

Fig. 2.3 The comparison between the diffusion-mediated (*left panel*) and the contact-based signaling processes (*right panel*). As shown in the *left panel*, the rate-limiting step determines the overall rate of the signaling pathway. In the contact-based signaling, the most compliant component in the signaling linkage determines the mechanics of the entire signaling pathway. Therefore, "gating spring" is mechanically the core of the contact-based signaling pathway. In this schematics, *r* denotes the reaction rate of each step, and *C* denotes the compliance of each structure (shown by the springs) in the signaling pathway

2.2.2 The Simple Mechanical Description

We now take a simple example to consider the mechanical signaling pathway more quantitatively. Assuming the external stimulus is a deformation (e.g., skin indentation caused by touch), D. This stimulus causes the conformational changes of the mechanotransduction channels (e.g., the deflection of the channel gate), Δd. As we just described, D is much larger than Δd, so there must be deformation (i.e., Δx) happened to the mechanical signaling pathway (Eq. 2.1):

$$D = \Delta d + \Delta x \tag{2.1}$$

If the signaling pathway that has an overall stiffness κ_{sp} and a channel gate that has an equivalent stiffness of κ_{ch}, the deformation D can be converted to a gating force, f_g.

The lumped compliance of the signaling pathway and the channel gate is:

$$C_{total} = C_{sp} + C_{ch} = \frac{1}{\kappa_{sp}} + \frac{1}{\kappa_{ch}} \tag{2.2}$$

$$f_g = \frac{D}{C_{total}} = \frac{\Delta x}{C_{sp}} = \frac{\Delta d}{C_{ch}} \tag{2.3}$$

Because D is much greater than Δd, we have an approximation for Eqs. 2.1 and 2.3:

$$D \approx \Delta x \tag{2.4}$$

$$f_g = \frac{D}{C_{sp}} \tag{2.5}$$

This is exactly what we meant that the compliance of the mechanical signaling pathway determines the gating forces applied onto the channel. Then, consider that the mechanical signaling pathway consists of a set of serially linked structures:

$$C_{sp} = C_1 + C_2 + \ldots + C_i + \ldots + C_n \tag{2.6}$$

Here, the subscripts denote all structures that form the force transmission chain (i.e., the mechanical signaling pathway). As we just described, C_{sp} represents the mechanical property of the formally defined "gating spring." In a special case when C_1 is much greater than the other elements, we could have another approximation:

$$C_{sp} \approx C_1 \tag{2.7}$$

$$f_g \approx \frac{D}{C_1} \tag{2.8}$$

In this case, the material/structure no.1 is sometimes called the "gating spring" as well even though it is just one of the components in the entire signaling pathway.

The gating of the mechanotransduction channel is controlled by the gating force, f_g:

$$\frac{P_o}{P_c} = \exp\left[-\left(\frac{\Delta\mu_0 - f_g.\Delta d}{k \cdot T}\right)\right] \tag{2.9}$$

P_o is the probability of the channel being in the opening state.
P_c is the probability of the channel being in the closed state.
$\Delta\mu_0$ is the difference in free energy between the closed and the open channel.
k is Boltzmann constant.
T is absolute temperature.

According to Eqs. 2.2, 2.3 and 2.4, we will have:

$$\Delta d = \frac{C_{ch}}{C_{ch} + C_{sp}} D \approx \frac{C_{ch}}{C_{sp}} D \qquad (2.10)$$

$$\frac{P_o}{P_c} = \exp\left[-\left(\frac{\Delta\mu_0 - \frac{C_{ch}}{C_{sp}^2} D^2}{k \cdot T}\right)\right] \qquad (2.11)$$

In Eq. 2.11, it is straightforward that the mechanosensory response of a given transduction channel, C_{ch}, to a given stimulus, D, is determined by the mechanical compliance provided by the mechanical signaling pathway, C_{sp}.

2.2.3 Model Predictions: Sensitivity and Dynamic Range

Sensitivity and dynamic range are two basic parameters to evaluate the performance of a sensory system. We will discuss the importance of "gating spring" in shaping these two parameters of a mechanosensory system.

Formally, sensitivity measures the system's response to a given signal (absolute sensitivity) or to the change in the signal intensity (relative sensitivity). For the simple model in Sect. 2.1.3, the absolute sensitivity measures the P_o at a given D:

$$P_o = \frac{\exp\left[-\left(\frac{\Delta\mu_0 - \frac{C_{ch}}{C_{sp}^2} D^2}{k \cdot T}\right)\right]}{1 + \exp\left[-\left(\frac{\Delta\mu_0 - \frac{C_{ch}}{C_{sp}^2} D^2}{k \cdot T}\right)\right]} \qquad (2.12)$$

In Eq. 2.12, a large C_{sp} leads to a smaller P_o. This means that for a given transduction channel, the system with a more compliant spring (i.e., force transmission pathway) responds less (i.e., smaller P_o) to the distal stimulus (D), namely, the absolute sensitivity is lower (Fig. 2.4, **upper panel**). We then consider the relative sensitivity, which measures how much decrease or increase in P_o will happen to the transduction channels for a change in the distal signal, ΔD, namely:

$$\text{Sensitivity}_{relative} = \frac{\Delta P_o}{\Delta D} \qquad (2.13)$$

Fig. 2.4 The compliance of the mechanical signaling pathway determines the sensitivity and the dynamic range of a mechanosensory system. *Upper* panel: the rigid system has a higher absolute sensitivity than the compliant system. *Middle* panel: the rigid system is also more sensitive to the changes in the signals (shown by the larger α_r in comparison to the α_c). *Lower* panel: the rigid system has a narrower dynamic range (shown by the *blue arrows*) than the compliant system (shown by the *red arrows*)

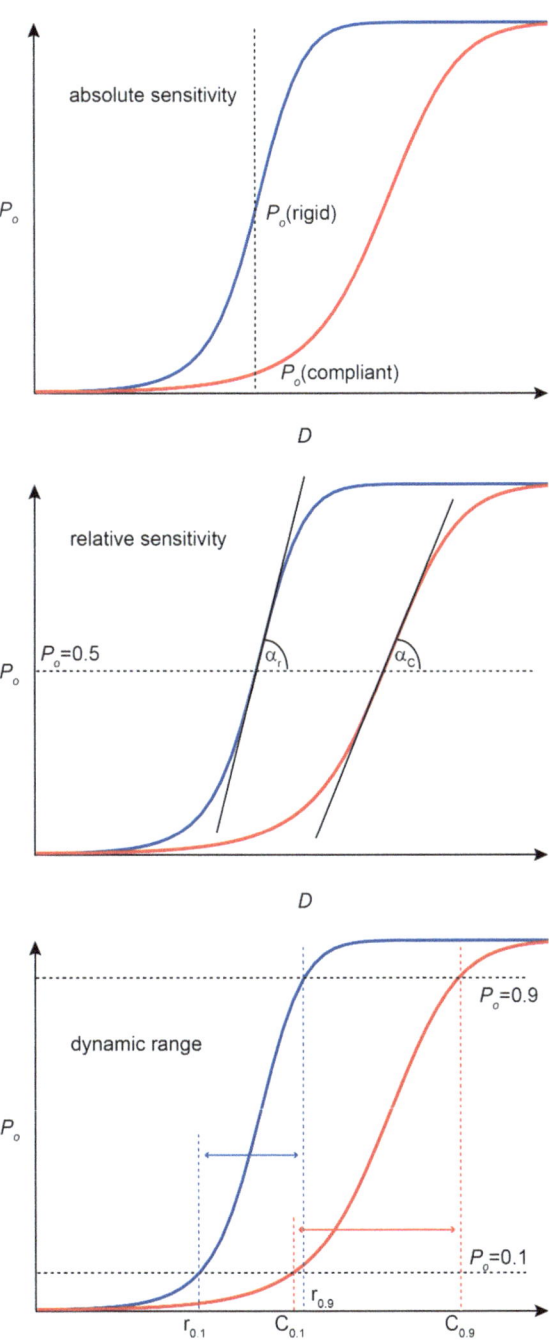

Therefore, the first derivative of P_o (Eq. 2.13) measures the relative sensitivity for a given mechanosensory system (i.e., a transduction channel with C_{ch} and a signaling pathway with C_{sp}). For simplicity, we first clean up Eq. 2.12 as the following:

$$\alpha = \frac{C_{ch}}{C_{sp}^{2}} \cdot \frac{1}{kT} \tag{2.14}$$

$$\beta = \exp\left(\frac{\Delta\mu_0}{k \cdot T}\right) \tag{2.15}$$

$$P_o = \frac{\exp\left(\alpha \cdot D^2\right)}{\beta + \exp\left(\alpha \cdot D^2\right)} \tag{2.16}$$

$$P_o' \frac{\exp\left(\alpha \cdot D^2\right)2\alpha\beta D}{\left[\beta + \exp\left(\alpha \cdot D^2\right)\right]^2} \tag{2.17}$$

We now analyze a special case of Eq. 2.17 when P_o is 0.5. In this case, we have:

$$\beta = \exp\left(\alpha \cdot D^2\right) \tag{2.18}$$

$$P_{o(P_o=0.5)}' = \frac{1}{2}\sqrt{\alpha \cdot \ln(\beta)} \tag{2.19}$$

Combine Eqs. 2.14 and 2.19, we obtain:

$$P_{o(P_o=0.5)}' = \frac{1}{2}\sqrt{\frac{C_{ch}}{C_{sp}^{2}} \cdot \frac{1}{kT} \cdot \ln(\beta)} = \frac{1}{2} \cdot \frac{1}{C_{sp}} \cdot \sqrt{\frac{C_{ch}}{kT} \cdot \ln(\beta)} \tag{2.20}$$

From Eq. 2.20, we see that the relative sensitivity of a mechanosensory system (at $P_o = 0.5$) depends on the compliance of the mechanical signaling pathway, C_{sp} (Fig. 2.4, **middle panel**). More specifically, when the force transmission pathway is more compliant, the system is less sensitive.

Combining the discussion on the absolute and relative sensitivities, we have an intuitive understanding on the role of C_{sp} in shaping system's sensitivity. When the materials between the distal stimulus and the transduction channel are compliant, more deformations will be taken up by the intermediate materials and less will happen to the transduction channels, so the opening probability of the transduction channel is small. In the extreme cases, if the intermediate material is infinitively compliant, then the channel will never open, and conversely, if the intermediate material is infinitively rigid, all the distal deformation will be directly transmitted to the channel, which most likely will overload the channel molecule or the membrane.

Then we consider the dynamic range of this sensory system. Dynamic range refers to the range of signals, within which the differences in the response of the sensory system reflect the changes in the incoming signals. When the incoming signal is weaker than the lower bound of this range, the system has no response to the signals. Conversely, when the signals are stronger than the upper bound of this range, the system always fully responds so that the output is saturated and cannot reflect the changes in the incoming signals.

For simplicity, we assume that if P_o is below 0.1, the system has "no response," while if P_o is above 0.9, the system appears to be "saturated." Now we consider the range of incoming stimuli, $[D_{0.1}, D_{0.9}]$, that the model system can resolve.

From Eq. 2.16, we obtain:

$$P_o = 0.1 \rightarrow \beta = 9 \cdot \exp\left(\alpha \cdot D_{0.1}^2\right) \rightarrow D_{0.1} = \sqrt{\frac{1}{\alpha}} \cdot \sqrt{\ln\left(\frac{\beta}{9}\right)} \qquad (2.21)$$

$$P_o = 0.9 \rightarrow \beta = \frac{1}{9} \cdot \exp\left(\alpha \cdot D_{0.9}^2\right) \rightarrow D_{0.9} = \sqrt{\frac{1}{\alpha}} \cdot \sqrt{\ln(9 \cdot \beta)} \qquad (2.22)$$

From Eqs. 2.22 and 2.14, we obtain:

$$\Delta D_{0.1-0.9} = D_{0.9} - D_{0.1} = C_{sp} \cdot \sqrt{\frac{kT}{C_{ch}}} \cdot \left[\sqrt{\ln(3 \cdot \beta)} - \sqrt{\ln\left(\frac{\beta}{3}\right)}\right] \qquad (2.23)$$

From Eq. 2.23, we can see that the compliance of the signaling pathway, C_{sp}, determines the dynamic range of the sensory system (Fig. 2.4, **lower panel**). More specifically, the increase in C_{sp} broadens the dynamic range (ΔD increases), namely, that the sensory system with a more compliant signaling pathway has a wider dynamic range in comparison to the system with a more rigid pathway.

As we see, the compliance of the mechanical signaling pathway shapes the performance of the sensory system. The increase in C_{sp} reduces the system's sensitivity but broadens the dynamic range and vice versa. On one hand, it is a trade-off for the system to have an optimal combination of the sensitivity and the dynamic range. On the other hand, it provides a certain flexibility for the sensory system to tune its signaling behavior by regulating the mechanical properties of the signaling pathways, so that one type of transduction channel could response to the mechanical stimuli that have different characteristics (e.g., intensities, frequencies, etc.) by coupling itself to a different gating spring. These calculations strengthen the point that the mechanical signaling pathway (or the "gating spring") is the core in mechanosensory transduction process.

2.2.4 Functional Implications

One important implication from the model is that the same transduction channel could be coupled to different signaling pathways in different sensory cells, as we have mentioned above. It is not yet clear whether this strategy is taken by the nature. One interesting phenomenon that might accord with this possibility is the observation of one type of mechanotransduction channel expressed in different mechanosensory cells that have different physiological functions. For example, NompC/TRPN, a fly mechanotransduction channel, is found in both type I and type II mechanoreceptors in flies (these receptors respond to different stimuli) (Liang et al. 2013; Yan et al. 2013; Zhang et al. 2013). These sensory cells have different tasks in physiology, so an interesting question is whether the mechanotransduction apparatuses in these cells provide different structural-mechanical supports for the gating of NompC. Another example is Piezo. Piezo has been found in many different types of cells, and these cells have differentiated functions in sensing mechanical signals (Honore et al. 2015; Volkers et al. 2015). Therefore, it is interesting whether Piezo also recruits different mechanical partners to shift its gating dynamics accordingly in different cells.

The second implication is the reference point that the transduction apparatus is tuned at the resting condition. As shown in Fig. 2.5, the stimulus-response curve has an S-shape, which indicates that at different points, the system has different relative sensitivities. When the system is tuned to point A at the resting condition, the system would mainly respond to the positive deformation but has a much smaller dynamic range for the negative deformations. Conversely, if the system is tuned to point C at the resting condition, the behavior of the system would be opposite. In comparison to point A and point C, if the system is tuned to point B (e.g., by applying positive pretension to the transduction channel), the system has an intermediate response at the resting state and can respond to both positive and negative signals. This means that the system can be both inhibited and excited. These three scenarios provide the engineering options for the mechanotransduction apparatus in which without changing any component in the transduction apparatus, the system's behavior can be tuned accordingly. It is also not clear whether these strategies are practically taken by any cells, but it could be an interesting point to look at in different mechanosensory cells.

2.2.5 Molecular Basis

The molecular basis of mechanosensory transduction is the least understood in all sensory processes; therefore, the molecular identities of the responder molecules and the gating springs remain elusive in nearly all model systems (probably except for the bacterial cell). In the bacterial cells, Mscs have been deeply studied, and it is clear that they respond to the force from the lipid bilayer (Kung et al. 2010).

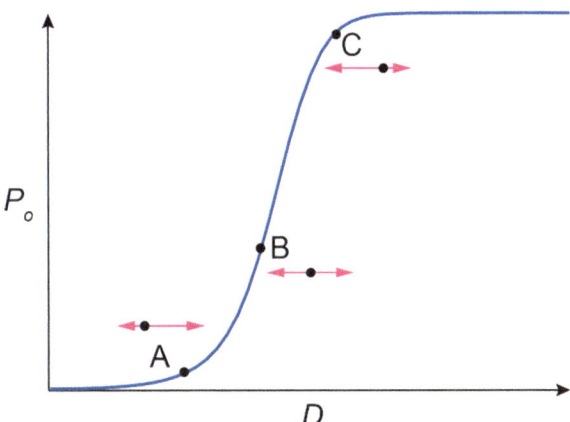

Fig. 2.5 The positions of the reference points of the system at the resting state determine the system's sensitivity and dynamic range in the sensory responses. At point A, the system would respond to a wide range of positive deformations. At point B, the system responds to both *negative* and *positive* deformations, while at point C, the system mainly responds to negative deformations

Therefore, the "force from the lipid" principle applies well to the Msc channels. In vertebrate hair cells, the tip links were considered as the compliant gating springs, but they were later shown to be much stiffer than what is expected to be the gating springs (Gillespie and Muller 2009). Therefore, the materials that provide the mechanical compliance in hair cells are still unknown. Several candidates, including the plasma membrane and the intracellular tether, were proposed but remain to be further demonstrated (Powers et al. 2012). Similarly, several candidate molecules have been proposed to be the components of the transduction channel complex, but it is not yet confirmed. In *C. elegans* touch receptors, it is clear that MEC-4/MED-10 complex functions as the responders (Goodman 2006). However, it is not clear where the mechanical compliance comes from. In fact, the molecular basis of the mechanotransduction apparatus in *C. elegans* touch receptors is one of the best studied model cells. It actually provides a good system to address the structural-mechanical basis of the transduction apparatus if the advantages in worm genetics and physiology are combined with the in situ nanomechanics measurements. *Drosophila melanogaster* is also a great system to address the molecular basis of the transduction apparatus, which we will elaborate in details in the following chapters.

References

Albert JT, Nadrowski B, Göpfert MC (2007) Mechanical signatures of transducer gating in the drosophila ear. Curr Biol 17:1000–1006

Corey DP, Hudspeth AJ (1979) Response latency of vertebrate hair cells. Biophys J 26:499–506

Gillespie PG, Muller U (2009) Mechanotransduction by hair cells: models, molecules, and mechanisms. Cell 139:33–44

Goodman MB (2006) Mechanosensation. WormBook:1–14

Honore E, Martins JR, Penton D, Patel A, Demolombe S (2015) The piezo mechanosensitive ion channels: may the force be with you! Rev Physiol Biochem Pharmacol 169:25–41

Howard J (2001) Mechanics of motor proteins and the cytoskeleton. Sinauer Associates, Publishers, Sunderland

Kung C, Martinac B, Sukharev S (2010) Mechanosensitive channels in microbes. Annu Rev Microbiol 64:313–329

Liang X, Madrid J, Gartner R, Verbavatz JM, Schiklenk C, Wilsch-Brauninger M, Bogdanova A, Stenger F, Voigt A, Howard J (2013) A NOMPC-dependent membrane-microtubule connector is a candidate for the gating spring in fly mechanoreceptors. Curr Biol 23:755–763

Liang X, Madrid J, Howard J (2014) The microtubule-based cytoskeleton is a component of a mechanical signaling pathway in fly campaniform receptors. Biophys J 107:2767–2774

Markin VS, Hudspeth AJ (1995) Gating-spring models of mechanoelectrical transduction by hair cells of the internal ear. Annu Rev Biophys Biomol Struct 24:59–83

O'Hagan R, Chalfie M, Goodman MB (2005) The MEC-4 DEG/ENaC channel of Caenorhabditis Elegans touch receptor neurons transduces mechanical signals. Nat Neurosci 8:43–50

Powers RJ, Roy S, Atilgan E, Brownell WE, Sun SX, Gillespie PG, Spector AA (2012) Stereocilia membrane deformation: implications for the gating spring and mechanotransduction channel. Biophys J 102:201–210

Volkers L, Mechioukhi Y, Coste B (2015) Piezo channels: from structure to function. Pflugers Arch 467:95–99

Walker RG, Willingham AT, Zuker CS (2000) A drosophila mechanosensory transduction channel. Science 287:2229–2234

Yan Z, Zhang W, He Y, Gorczyca D, Xiang Y, Cheng LE, Meltzer S, Jan LY, Jan YN (2013) Drosophila NOMPC is a mechanotransduction channel subunit for gentle-touch sensation. Nature 493:221–225

Zhang W, Yan Z, Jan LY, Jan YN (2013) Sound response mediated by the TRP channels NOMPC, NANCHUNG, and INACTIVE in chordotonal organs of Drosophila larvae. Proc Natl Acad Sci U S A 110:13612–13617

Chapter 3
Mechanoreceptors in *Drosophila melanogaster*

Abstract In Chaps. 1 and 2, we introduced the physiological roles of mechano-transduction and the conceptual design of a mechanoreceptor. How does a real mechanoreceptor look like and what are its specific functions in physiology? In this chapter, we provide the introduction for the specialized mechanoreceptors in *Drosophila melanogaster*, including their distributions, tissue/cellular organizations, and physiological functions. As shown in this chapter, the mechanosensory transduction is a fundamental process in the daily life of a fly, for example, in walking, standing, flying, crawling, jumping, hearing, defecation, mating, etc. The broad involvement of mechanotransduction in fly's life makes the fly mechanoreceptors great models to study the molecular, structural, and mechanical basis of mechanotransduction.

3.1 Overview of Fly Mechanoreceptors

Fly has two major types of mechanoreceptors, type I and type II (Singhania and Grueber 2014; Lumpkin et al. 2010; Gillespie and Walker 2001). Type I mechanoreceptors contain the ciliated sensory neurons, and type II mechanoreceptors are multidendritic neurons (i.e., non-ciliated) (Table 3.1). These mechanoreceptors are important in the daily life of a fly, including walking, flying, mating, etc. In Box 3.1, we provide four examples, two in adult files and another two in fly larvae, to show how these mechanoreceptors contribute to fly physiology.

3.2 Type I Mechanoreceptors

Type I mechanoreceptors include the external sensory (ES) organs and the chordotonal organs (CHO) (Table 3.1). External sensory organs include the bristle sensillum that has an external hair structure and the campaniform sensillum that has an external dome-like cuticle structure (Keil 1997). CHO cells do not have the external structures and are usually embedded under the cuticle structures (Keil 1997). The

© The Author(s) 2017   27
X. Liang et al., *Mechanosensory Transduction in Drosophila Melanogaster*,
SpringerBriefs in Biochemistry and Molecular Biology,
DOI 10.1007/978-981-10-6526-2_3

Table 3.1 Fly mechanoreceptors

Mechanoreceptor			Localization	Function
Type I (ciliated)	External sensory organ	Bristle sensilla	Whole body	Direct touch
		Campaniform sensilla	Joints, wing base, haltere base, leg	Cuticle deformation
		Larval sensory organs	Larval body wall	Possibly tactile
	Chordotonal organ (i.e., scolopidial organ)		Joints, antenna, leg, wing, haltere, larval body wall	Mechanical stress
Type II (multidendritic)	Dendritic arborization (da) neuron	Class I	Larval body wall	Proprioception
		Class II		Possibly gentle touch
		Class III		Gentle touch
		Class IV		Harsh touch
	Tracheal dendritic (td) neuron Ghysen and Dambly-Chaudiere (1993), Bodmer and Jan (1987)		Tracheal branches	Possibly proprioception
	Bipolar dendritic (bd) neuron Suslak et al. (2015), Cheng et al. (2010)		Larval body wall	Stretch, proprioception

common feature of the type I mechanoreceptors is that they all contain a ciliated sensory neuron and a set of supporting cells (Singhania and Grueber 2014; Keil 1997), so they are also referred to as the ciliated mechanoreceptors.

3.2.1 Bristle Sensilla

Bristles fully cover the fly body. The larger and stouter bristles are called macro-chaetes and smaller ones are microchaetes. Macrochaetes often localize at a few fixed positions, while the microchaetes are nearly everywhere covering the fly body (Fig. 3.1). For example, there are macrochaetes and microchaetes on the thorax, head, and abdomen; there are also bristles on fly antenna, legs, and around the neck region (Fig. 3.1a). These bristle receptors sense the direct touch or more precisely, the deflection of the hairlike structures caused by the environmental forces. One example is that the recurved bristles on the margin of fly wings participate the mechanosensory process in the defensive kicking behavior when the parasitic mites are invading over the wing margin (thereby may touch these bristles) (Li et al. 2016).

Box 3.1: Are the Mechanoreceptors Dispensable in Fly's Life?

Flight stability (Bechstedt et al. 2010; Fox and Daniel 2008; Fayyazuddin and Dickinson 1996): flies need to fly. In order to keep its body balanced when flying, the mechanoreceptors at fly wings and halteres provide the sensory feedback inputs that feed into the motor neurons and the central nervous system (CNS). By doing so, flies are able to instantly adjust the flying motions and keep their bodies balanced during the flying maneuvers.

Courtship song (Boekhoff-Falk and Eberl 2014; von Philipsborn et al. 2011; Dornan and Goodwin 2008): male flies play their "love song" to attract the female flies to start the copulation. The courtship song of the male fly is generated by a set of intricate behaviors, including following, tapping, licking the female, and finally vibrating the wing to generate the male courtship song. Mechanoreceptors in the wings and legs contribute to these processes. On the other hand, female flies hear the courtship song using the auditory organs (the "Johnston's organ") in the antenna. Just like in human beings, mechanotransduction is the cell biological basis of hearing in fly.

Avoidance of the natural enemy (Robertson et al. 2013): fly larvae are often attacked by the predators. For example, wasps attack fly larvae using the sharp ovipositor and lay eggs in the bodies of fly larvae. Larvae avoid this attack by the nocifensive escape locomotion, for example, rolling their bodies. These behaviors are neurologically initiated by the mechanoreceptors embedded in the larval epidermis.

Larvae defecation (Zhang et al. 2014): the defecation behavior of fly larvae involves the sensing of the contraction/stretch status of the anal muscle. This is achieved by the mechanosensory neurons around the anal slit. The loss of these sensory neurons leads to the abnormal contraction of the anal muscles and thereby defective behaviors in defecation.

Each mechanosensitive bristle sensillum contains a ciliated sensory neuron (Fig. 3.1b, c). It is a bipolar neuron that has an unbranched dendrite, a soma, and an axon (Fig. 3.1b, c). The distal part (2–3 μm) of the neuronal dendrite contains a modified cilium that origins from the basal body (Fig. 3.1). The dendritic region on the apical side of the cilium is termed as the "outer segment," while the region on the basal side of the cilium is termed as the "inner segment." In the distal end of the cilium, the "9 + 0" axonemal structure elaborates into a specialized microtubule-based structure that supports a dilatory body in the dendrite, i.e., the tubular body (Fig. 3.1). The distal tip of the tubular body makes a direct contact with the base of the hairlike structure and thereby is at the position to receive the mechanical stimuli when the hair is deflected. Therefore, this distal tip is thought to be the site of mechanotransduction and to enclose the mechanosensory apparatus that transduces the mechanical stimuli into the intracellular signals. The molecular basis of mechanotransduction in bristle cells is still largely unknown, which will be further discussed in Chap. 4.

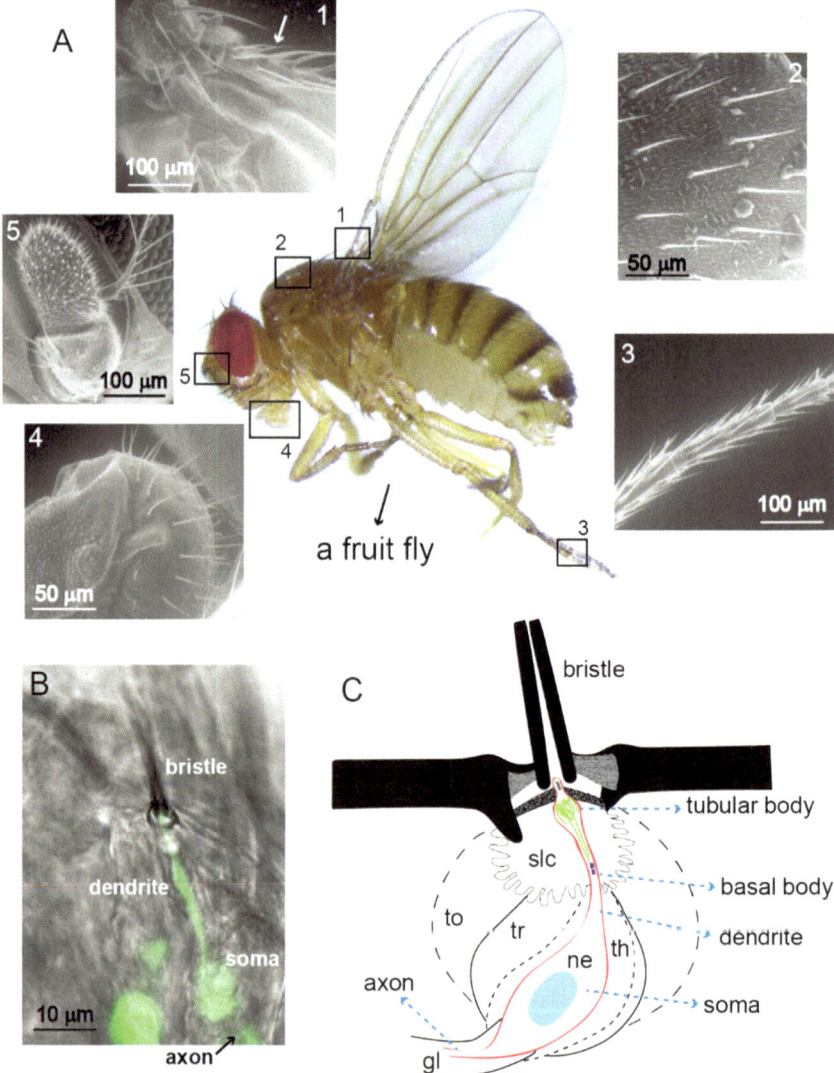

Fig. 3.1 Bristle sensillum. (**a**) The images of a fruit fly and the bristle sensilla. Images 1–5 are the scanning electron microscope images of the bristles at different locations in the fly body. Image 1, wing margin bristles; image 2, thorax bristles; image 3, leg bristles; image 4, labella bristles; image 5, antenna bristles. (**b**) A fluorescent image of the sensory neuron (a GFP-labeled neuron) in the bristle sensillum (transmission light image). (**c**) The cellular organization of the bristle sensillum in which the sensory neuron (*red outlined*) is surrounded by a set of supporting cells (Keil 1997), i.e., *ne* neuron, *th* thecogen, *tr* trichogen, *to* tormogen, *slc* sensillum lymph cavity. Slc is filled with the K+-rich receptor lymph. The different parts of the sensory neuron are indicated by the *blue dashed arrows* (The schematic is modified after the reference (Keil 1997))

Surrounding this ciliated sensory neuron is a set of supporting cells. They are the thecogen cell, tormogen cell, and trichogen cell (Fig. 3.1). These cells are of the same lineage as the sensory neuron, namely, that these three supporting cells and the sensory neuron are differentiated from the same ancestor, namely, the sensory organ precursor (SOP) cell (Singhania and Grueber 2014; Keil 1997). The SOP cell originates in the epidermis and is committed the neurogenic fate. The first division of the SOP cell gives rise to two secondary precursor cells: the pIIa (more basal) and pIIb (more apical) cells. In the second division, pIIa cell yields trichogen and tormogen cells, while pIIb cell gives a small glial cell and a tertiary precursor cell, pIIIb. In the third division, the pIIIb cell divides into the sensory cell and the thecogen cell (Singhania and Grueber 2014; Keil 1997).

The supporting cells are essential parts of the bristle sensillum and have specialized supporting functions (Keil 1997). During the development, the sensory neuron sinks to the basalmost position and forms two processes: the axon to the central nervous system and the dendrite toward the apical surface of the epidermis. Thecogen cell forms a glia-like envelop around the dendritic inner segment of the sensory neuron. It also secretes the extracellular matrix that forms the dendritic sheath covering the dendritic outer segment. The soma and axon of the sensory neuron are enveloped by a glial cell that is possibly of the epithelial lineage (Fig. 3.1). The trichogen cell and tormogen cell are around the thecogen cell and form the surrounding structures including the outer sensillum lymph space and the socket septum. The sensillum lymph space contains the receptor lymph that serves as the extracellular fluid of the sensory cilium. This receptor lymph is rich in K^+ which differs from the composition of the Na^+-rich hemolymph. The high K^+ in receptor lymph is kept by an ATP-dependent apparatus that transports the K^+-ions against the concentration gradient. The movement of K^+ creates an electrical activity can be recorded as the "transepithelial voltage" (TEV) in the bristle sensilla.

3.2.2 Campaniform Sensilla

Campaniform sensilla are only found in insects so far (Keil 1997). They localize at the joints between the different body parts, for example, between the leg segments (Fig. 3.2). They sense the cuticle deformation when the flies are crawling, walking, flying, and jumping. There are small fields of campaniform sensilla at the fly wing base (Fig. 3.2). These receptors report the mechanical signals that arise from the wing beats. The biggest fields of campaniform sensilla are on the fly halteres (Fig. 3.2). These campaniform receptors are extremely important for the flight behavior of flies. Briefly, flies are extremely good flyers, for example, their forewings can perform very rapid maneuvers in air. However, the stable flying behavior is only possible if the maneuvers are reported to the central nervous system and the motor systems as fast as the maneuvers themselves; otherwise flies would lose the body balance. Therefore, flies require a gyroscope, which is often used in ships and aircrafts, to keep the body steady. Halteres act as the gyroscopes for the flies. They

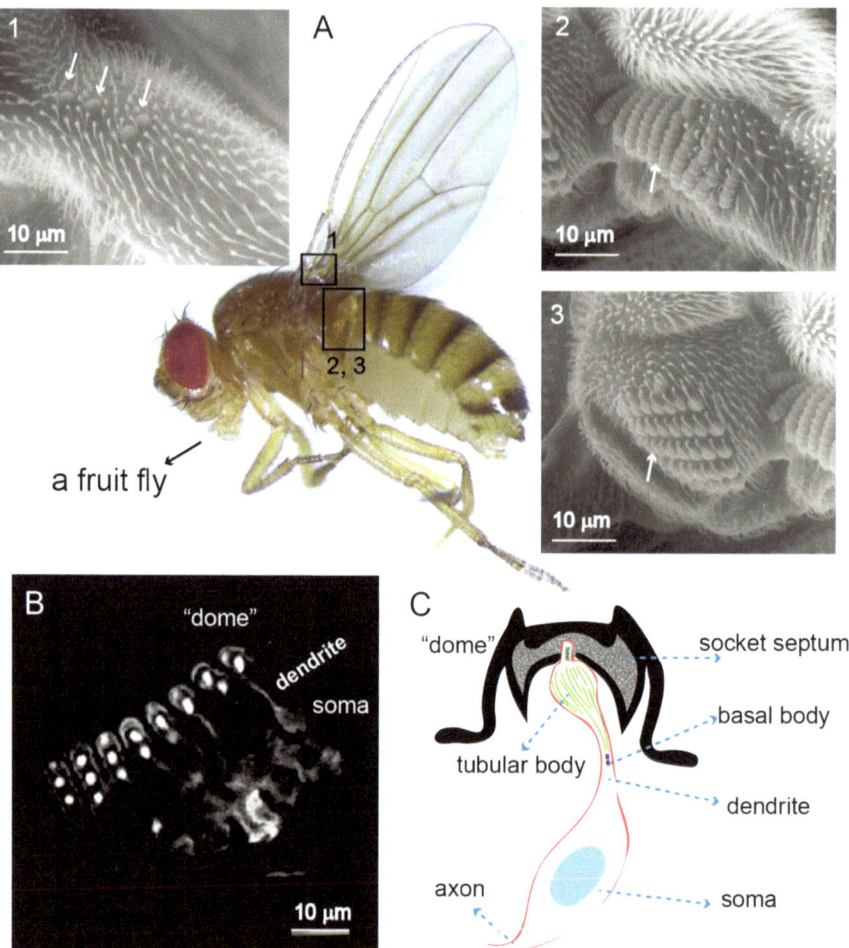

Fig. 3.2 Campaniform sensilla. (**a**) The images of a fruit fly and the campaniform sensilla. Images 1–3 are the scanning electron microscope images of the campaniform sensilla at the wing and haltere. Image 1, wing campaniform receptors; image 2, haltere campaniform receptors (pedicel); image 3, haltere campaniform receptors (scabellum). Note that the campaniform receptors at different locations (e.g., wing or haltere) may have different external cuticle structures, as indicated by the *white arrows* in each image. (**b**) A fluorescent image of a campaniform receptor field in the haltere pedicel (GFP labeled neurons). (**c**) The cell structure of the sensory neuron in the campaniform receptors. The supporting cells are omitted in this schematic, but they are arranged in a similar pattern as in the bristle sensillum (Fig. 3.1c). The different parts of the sensory neuron are indicated by the *blue dashed arrows*

are the degenerated hindwings. Halteres are oscillating in the same frequency as the forewing beats but in the antiphase manner. When flies initiate a maneuver, the halteres are deflected in relative to the fly bodies. Such deflections stretch or compress the cuticle structures at the base of the haltere. There are three fields of about 100 campaniform receptors at the base of the haltere (Fig. 3.2). These receptors are excited or inhibited by the cuticle stretch or compression and thereby provide rapid sensory feedback inputs for the CNS and motor system to adjust the body status and wing beats accordingly.

Similar to the bristle sensillum, each campaniform sensillum contains one sensory cell (Fig. 3.2). The overall sensillar structure is similar to the bristle sensory cells. The major difference between the bristle and campaniform sensory cell is the outer dendritic segment. This is thought to be mainly due to the completely different external cuticle structures in these two types of cells (Figs. 3.1 and 3.2). Therefore, the outer segments, which are associated with the cuticle, have different geometries and intracellular structures. The outer segment of the sensory cell in the haltere campaniform sensillum has been carefully studied. On the distal side of the axonemal structure, the ciliary microtubules elaborated into specialized microtubule-based structure, i.e., the tubular body (Fig. 3.2). On the distal side of the tubular body, i.e., the distal tip of the dendrite, there exists a specialized compartment that abuts the cuticle and contains a highly ordered microtubule-based intracellular structure. Like in the bristle receptors, this dendritic tip is thought to be the site of mechanotransduction, namely, the place where the deformation of the haltere cuticle is converted to the intracellular signals. We will elaborate on the ultrastructure of this mechanotransduction site in the next section when we discuss the mechanotransduction mechanism of the campaniform receptor.

The sensillar organization of the campaniform sensillum is similar to the bristle sensillum (Fig. 3.1). The sensory cell is surrounded by the three supporting cells, i.e., the thecogen cell, tormogen cell, and trichogen cell. The receptor lymph is also rich in K^+. Its major characteristics is the dome-like cuticle structure that overlays the dendritic tip of the sensory cells. This dome-like structure is thought to function as a mechanical amplifier that sensitizes the campaniform sensilla (Menon et al. 2009).

3.2.3   Chordotonal Organ

Chordotonal organs, also called the scolopidial organs, are the homologues of the external sensory organs (Keil 1997). The sensillar organization and the development process of the scolopidial organs are similar to the bristle and campaniform sensilla. Scolopidial organs have no external cuticle structures (Fig. 3.3a, b). Instead, they are inside the body and are sensors for the mechanical stress (Fig. 3.3a). In most cases, a scolopidial unit contains two sensory cells (Fig. 3.3c) (there could be more than two sensory cells in some cases). The cellular organization of the scolopidial sensory cell is generally similar to the sensory cells in bristle and campaniform receptors. The scolopidial sensory cell is a bipolar neuron and contains a cilium structure in the dendrite (Fig. 3.3b). This cilium has the "9 + 0" axonemal structure but much longer than the one in the bristle and campaniform receptor. The basal body localizes at nearly the middle of the dendrite (Fig. 3.3c), which separates the

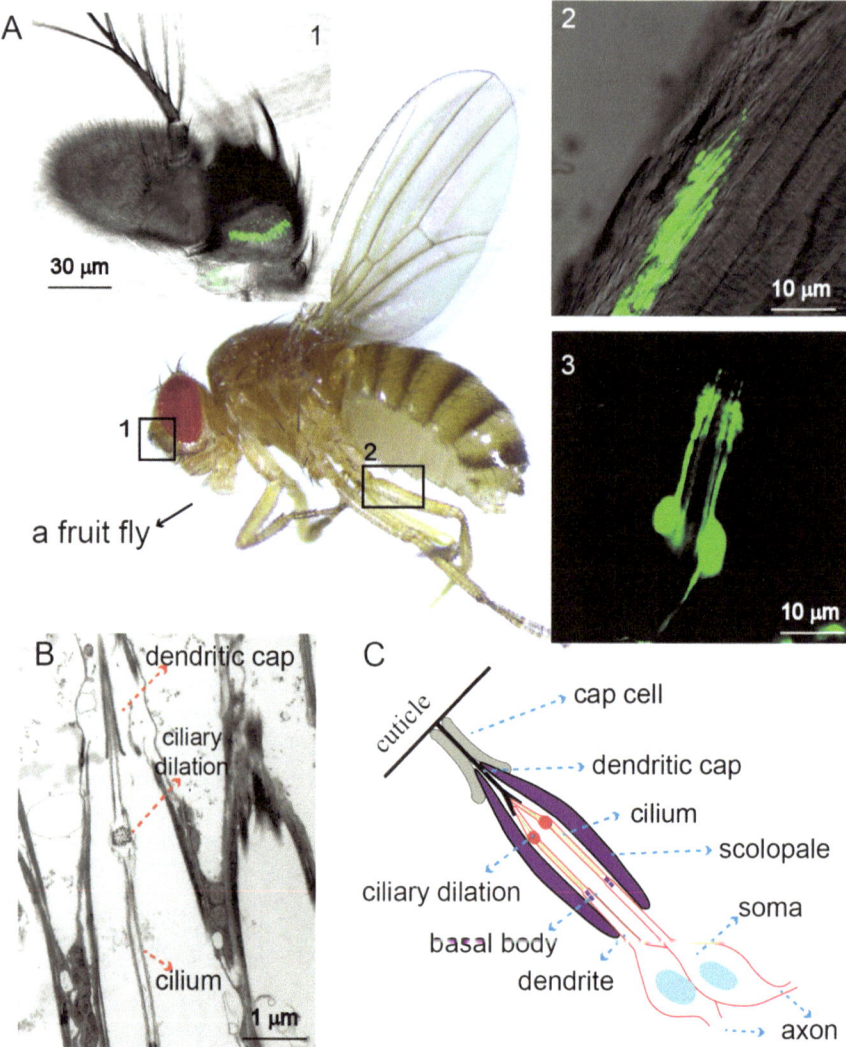

Fig. 3.3 Scolopidial cells. (**a**) The images of a fruit fly and the scolopidial cells. Images 1–3 are confocal microscope images of the scolopidial cells (chordotonal organs). Image 1, the Johnston's organ in a fly antenna; image 2, the femoral chordotonal organ (leg); image 3, a larval chordotonal organ (not indicated in the adult fly image). Note that the scolopidial cells at different locations usually cluster into a sensory organ and they may differ in their sizes (i.e., lengths). (**b**) A transmission electron microscope image of the ciliary part in a scolopidial cell in a fly antenna. The dendritic cap, ciliary dilation, and the cilium are indicated with the *red dashed arrows*. (**c**) The cellular structure of a scolopidial unit, in which the different parts of the scolopidial unit are indicated by the *blue dashed arrows*

dendrites into the proximal and distal segments. The proximal part of the dendrite (inner segment) contains the ciliary rootlet. The distal part of the dendrite (outer segment) is the ciliary structure (Fig. 3.3b, c). This ciliary structure is modified at about one third of its length to the distal tip to form a dilatory structure, named the ciliary dilation (Fig. 3.3c). In the ciliary dilation, there exists the electron dense structure that has a form of paracrystalline and is made of unknown substance (Fig. 3.3b). Interestingly, the microtubules in the cilium are bent in the ciliary dilation region and reform the ciliary structure on the distal side of the ciliary dilation (Bechstedt et al. 2010). On the distal side of the dilation, the dendritic tip terminates in the dendritic cap (Fig. 3.3b), an extracellular sheathlike structure. This dendritic cap is enveloped by an attachment cell that finally links the scolopidial sensory cell to the cuticle (Fig. 3.3c). The dendritic tip is thought to be the site of mechanotransduction in this type of sensory cell and encloses the molecular apparatus that transduces the mechanical signals coming from the cuticle through the attachment cell and the dendritic cap (Boekhoff-Falk and Eberl 2014; Bokolia and Mishra 2015).

The largest scolopidial sensory organ in flies is the Johnston's organ (Boekhoff-Falk and Eberl 2014; Bokolia and Mishra 2015). It is inside the second segment of the antenna (Fig. 3.3a). This organ contains more than 200 scolopidial units and nearly 500 scolopidial sensory cells (Kamikouchi et al. 2009). The cross section of the second segment of the antenna shows that these scolopidial units have a fan-shaped organization (Fig. 3.3a). The distal ends of all the scolopidial cells converge and attach to a stalk structure at the joint region between the second and third segments of the antenna. Johnston's organ is the auditory organ of fly, but it also contributes to fly's perception of gravity and wind (Kamikouchi et al. 2009; Yorozu et al. 2009). The mechanical principles and the working mechanisms of the Johnston's organ will be discussed in Chap. 4.

3.2.4 Ciliated Mechanoreceptors in Fly Larvae

There are ciliated mechanoreceptors in fly larvae. These sensory organs are thought to be homologues of the type I mechanoreceptors in the adult flies (Fig. 3.4). Most of these sensory organs are provided with the larval body wall and localize at the thoracic and abdominal segments. Although the exact functions of these sensory cells haven't been definitively characterized, some of these sensory cells are thought to be the tactile receptors, while others are likely chemoreceptors (Demerec 2008). The external cuticle structures of these cells take the form of hairs, different types of bristles, and different types of cones (Demerec 2008). The development and the sensillar organization of these sensory organs are thought to be similar to the adult external mechanosensory cells.

There are also scolopidial organs in fly larvae (Fig. 3.3a). One type of the chordotonal organs is called the scolopophorus organs. They are compound sensory organs placed between the body wall and the muscle layer but are attached at both ends to the hypoderm. These organs sense the mechanical stress and function pri-

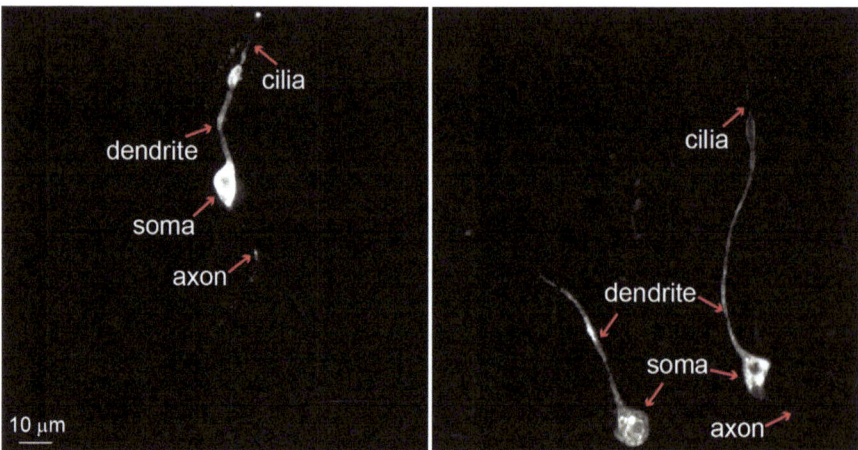

Fig. 3.4 Larval ciliated sensory cells. (*Left*) A ciliated sensory cell in the anterior part of the fly larva. (*Right*) A pair of ciliated sensory cells in the posterior part of the fly larva. These neurons are ciliated sensory cells that may detect the chemical or mechanical signals

marily to control the muscle tension. They are present in all the larval body segments and are degenerated during the pupation stage. The other type of the chordotonal organs in fly larvae can be excited by the vibration or sound (Zhang et al. 2013), which allows the larvae to respond to the predators, for example, the wasps and jackets, at a distance. These cells are also found to provide the sensory input for the proprioception of the fly larvae. The cell organizations of these scolopidial cells are generally similar to scolopidial cells in the adult flies and are primarily different in their sizes (e.g., dendritic lengths), probably to match their different functions.

3.3   Type II Mechanoreceptors

The type II sensory cells in fly are the non-ciliated and multidendritic (md) neurons, including the tracheal dendrite neurons (td), bipolar dendrite neurons (bd), and dendritic arborization (da) neurons (Singhania and Grueber 2014; Grueber et al. 2002). The class I, II, III, and IV type da neurons are among the best studied md neurons. In contrast to the type I mechanosensory organs, type II mechanoreceptors have no specialized supporting cells. These neurons are present in all body segments and placed between the epidermis and a layer of extracellular matrix. They have elaborated branching dendritic structures and a single axon that feeds into the CNS. All four types of da neurons have been shown to be involved in the mechanosensory processes (Yan et al. 2013; Robertson et al. 2013; Tsubouchi et al. 2012; Cheng et al. 2010).

Fig. 3.5 A class I da neuron. The boundaries (*bottom left* and *top right corners* in the image) between the larval body segments are visible in the image due to the autofluorescence of the cuticle

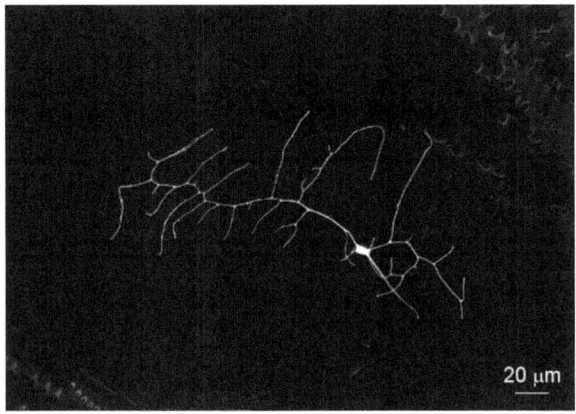

20 µm

3.3.1 Class I da Neuron

Among all four types of da neurons, class I neuron has the simplest dendritic morphology (Fig. 3.5). In each hemisegment, three class I neurons, named as vpda, ddaD, and ddaE, innervated both the dorsal and the ventral (only limited) region of the fly larval body wall (Grueber et al. 2002). In three class I neurons, vpda has one primary dendrite that branches in a nearly regular manner along its length into the secondary branches. Slightly different from the vpda neuron, ddaE and ddaD have the fan-shaped dendritic morphologies, and their dendrites project anteriorly and posteriorly, respectively. Despite their smooth and low complexity morphologies, class I neurons were shown to be important in the fly larval proprioception (Cheng et al. 2010). Proprioception refers the sensory feedback process which animals use to monitor the status of their body parts/segments in order to coordinate their bodies for the balance and locomotion. The experimental evidence is that the mutant larvae, in which the function of class I neurons is lost, showed the clear defects in locomotion (Cheng et al. 2010). Therefore, the localization of class I neurons in the body segments, their association with the larval body wall, their dendritic morphology, and their relative positions to the muscles suggest that the class I neurons are likely the stretch-sensitive receptors that provide sensory inputs in the neuronal circuitry of the proprioception. Note that it was also reported that class I neuron is not required in the gentle touch response of the fly larvae (Tsubouchi et al. 2012).

3.3.2 Class II da Neuron

Class II da neurons also have relatively simple morphologies (Fig. 3.6) (Grueber et al. 2002). There are four members in this family, i.e., vda A, vdaC, ldaA, and ddaB. They cover mostly the ventral region and some parts of the lateral and dorsal regions. Despite being similarly simple as the class I neurons, the morphological

Fig. 3.6 Two class II da
neurons (indicated by two
smaller arrows) and one
class I neuron (indicated
by a *bigger arrow*). The
class I and class II neurons
can be clearly
distinguished from the
differences in their
morphological patterns

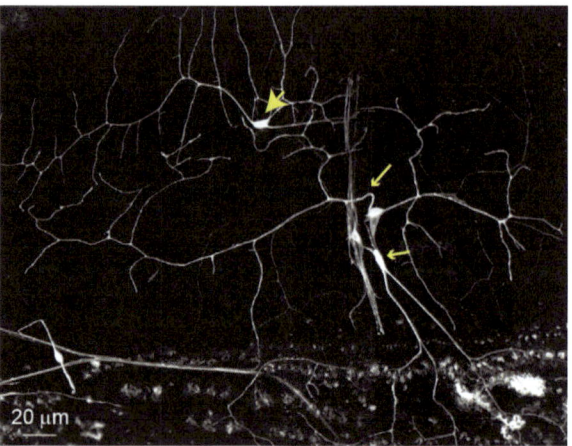

pattern of the dendrites in the class II neurons is different from the class I neuron.
Class II neurons have more than one primary branches that are long but less branched
as they extend to the distant regions. There are only a few studies on the function of
the class II neurons, probably due to the lack of a specific gal4 driver. A recent study
reported that class II neurons, together with the class III neurons, are involved into
fly larvae's perception of gentle touch (Tsubouchi et al. 2012).

3.3.3 Class III da Neuron

Class III da neurons include five members, i.e., ddaF, ddaA, ldaB, v'pda, and
vdaD. Compared to the class I and class II neurons, class III neurons have more
extensively branched dendritic structures (Fig. 3.7) (Grueber et al. 2002). Therefore,
five class III da neurons together nearly tile the entire abdominal hemisegment
(~70%). The most striking feature of the class III neurons is the spike-like protrusions that are along the primary and secondary branches (Fig. 3.7), which clearly
distinguishes the class III neurons from the other da neurons. Class III neurons were
reported as the sensor for the gentle touch stimuli in fly larvae. The key evidence
include (1) when the neuronal activities are silenced by overexpressing the tetanus
toxin light chain (UAS-TNT-E) in class III neuron, fly larvae's behavior responses
to the gentle touches are significantly reduced (Tsubouchi et al. 2012); (2) individual class III neurons have a clear response to the touch stimuli as shown by using the
fluorescent calcium imaging and the electrophysiological recording methods (Yan
et al. 2013); and (3) the number of the spike-like protrusions in the class III neurons,
likely acting as the harbor of molecule sensors, correlates with the strength of touch
responses (Tsubouchi et al. 2012).

Fig. 3.7 Two class III da neurons (indicated by two *arrows*). Class III neurons cover most of the larval body wall and are characterized by the spike-like protrusions extended from the primary and secondary branches

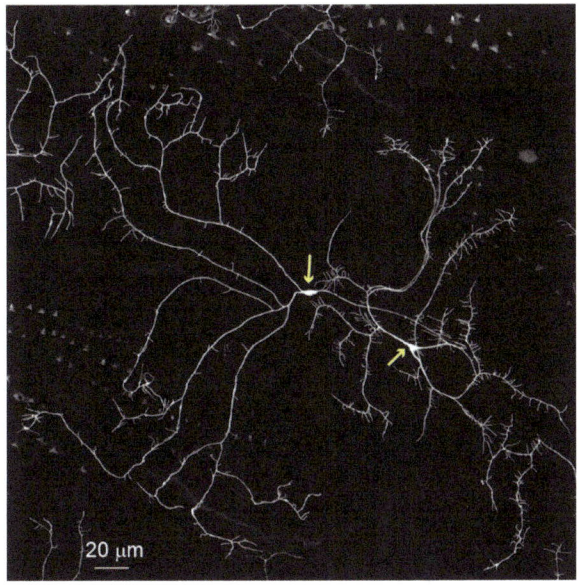

3.3.4   Class IV da Neuron

Among all the da neurons, class IV type is the best studied (Jan and Jan 2010). Its highly branched and complex morphology (Fig. 3.8) makes it a great model to study the cellular and molecular mechanisms underlying dendritic morphogenesis (Jan and Jan 2010; Grueber et al. 2002). There are three members of class IV neurons, i.e., vdaB, v'ada, and ddaC. The highly complex branching dendrites of these neurons nearly fill the entire larval body wall (~ 100%) with arbors (Grueber et al. 2002). Morphological analysis shows that each matured class IV neuron has more than a thousand dendritic branches and more than six branch orders (based on the Strahler analysis), which clearly distinguishes the class IV neurons from the other da neurons. Interestingly, class IV neurons are found to take part in several sensory processes, including mechanosensation, thermoception, as well as the light perception (Robertson et al. 2013; Kim et al. 2012; Xiang et al. 2010; Hwang et al. 2007). As a mechanoreceptor, the class IV neuron was reported to be a harsh touch sensor, in contrast to the class III neuron that acts as the gentle touch sensor. The key evidence are (1) class IV neurons are necessary for larvae's escape behavior in response to the predator's mechanical attack (Robertson et al. 2013; Hwang et al. 2007); (2) key mechanosensitive ion channels and other mechanotransduction-associated channel subunits, e.g., DmPiezo and Ppk, express in class IV neurons (Guo et al. 2014; Gorczyca et al. 2014; Kim et al. 2012); and (3) class IV neurons do not

Fig. 3.8 A class IV da
neuron. Class IV neurons
completely cover the larval
body wall and are
characterized by their
extensively branched
dendritic morphologies

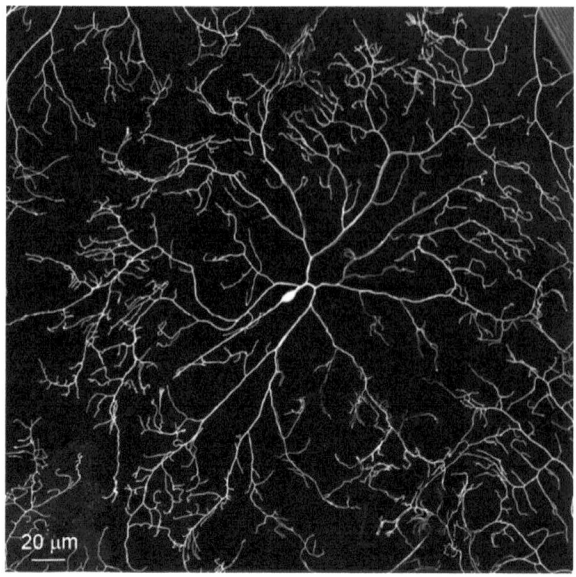

respond to the gentle touch stimuli (Yan et al. 2013; Tsubouchi et al. 2012). However,
it has not been experimentally demonstrated by in vivo calcium imaging and electrophysiological recording, as in the case of class III neurons, that the class IV
neurons can be excited directly for force, rather than other physical or chemical
factors secondary to the force stimuli.

3.4 Summary

Type I and type II mechanoreceptors together allow the flies (adults and larvae) to
receive various types of mechanical information from their environments. Although
many of these receptors remain to be further characterized, some of these receptors
have been studied extensively. In the next chapters, we will discuss in detail the
mechanotransduction in these cells.

References

Bechstedt S, Albert JT, Kreil DP, Muller-Reichert T, Gopfert MC, Howard J (2010) A doublecortin
    containing microtubule-associated protein is implicated in mechanotransduction in Drosophila
    sensory cilia. Nat Commun 1:11. https://doi.org/10.1038/ncomms1007
Bodmer R, Jan YN (1987) Morphological differentiation of the embryonic peripheral neurons in
    Drosophila. Roux Arch Dev Biol 196(2):69–77. https://doi.org/10.1007/BF00402027

Boekhoff-Falk G, Eberl DF (2014) The Drosophila auditory system. Wiley Interdiscip Rev Dev Biol 3(2):179–191. https://URL.org/10.1002/wdev.128

Bokolia NP, Mishra M (2015) Hearing molecules, mechanism and transportation: modeled in Drosophila melanogaster. Dev Neurobiol 75(2):109–130. https://doi.org/10.1002/dneu.22221

Cheng LE, Song W, Looger LL, Jan LY, Jan YN (2010) The role of the TRP channel NompC in Drosophila larval and adult locomotion. Neuron 67(3):373–380. https://doi.org/10.1016/j.neuron.2010.07.004

Demerec M (2008) Biology of drosophila. Cold Spring Harbor Laboratory Press, Woodbury

Dornan AJ, Goodwin SF (2008) Fly courtship song: triggering the light fantastic. Cell 133(2):210–212. https://doi.org/10.1016/j.cell.2008.04.008

Fayyazuddin A, Dickinson MH (1996) Haltere afferents provide direct, electrotonic input to a steering motor neuron in the blowfly. Calliphora J Neurosci 16(16):5225–5232

Fox JL, Daniel TL (2008) A neural basis for gyroscopic force measurement in the halteres of Holorusia. J Comp Physiol A Neuroethol Sens Neural Behav Physiol 194(10):887–897. https://doi.org/10.1007/s00359-008-0361-z

Ghysen A, Dambly-Chaudiere C (1993) The specification of sensory neuron identity in Drosophila. BioEssays 15(5):293–298. https://doi.org/10.1002/bies.950150502

Gillespie PG, Walker RG (2001) Molecular basis of mechanosensory transduction. Nature 413(6852):194–202. https://doi.org/10.1038/35093011

Gorczyca David A, Younger S, Meltzer S, Kim Sung E, Cheng L, Song W et al (2014) Identification of Ppk26, a DEG/ENaC channel functioning with Ppk1 in a mutually dependent manner to guide locomotion behavior in drosophila. Cell Rep 9(4):1446–1458. http://dx.doi.org/10.1016/j.celrep.2014.10.034

Grueber WB, Jan LY, Jan YN (2002) Tiling of the Drosophila epidermis by multidendritic sensory neurons. Development 129(12):2867–2878

Guo Y, Wang Y, Wang Q, Wang Z (2014) The role of PPK26 in drosophila larval mechanical nociception. Cell Rep 9(4):1183–1190. doi:http://dx.doi.org/10.1016/j.celrep.2014.10.020

Hwang RY, Zhong L, Xu Y, Johnson T, Zhang F, Deisseroth K et al (2007) Nociceptive neurons protect Drosophila larvae from parasitoid wasps. Curr Biol 17(24):2105–2116. https://doi.org/10.1016/j.cub.2007.11.029

Jan YN, Jan LY (2010) Branching out: mechanisms of dendritic arborization. Nat Rev Neurosci 11(5):316–328. https://doi.org/10.1038/nrn2836

Kamikouchi A, Inagaki HK, Effertz T, Hendrich O, Fiala A, Gopfert MC et al (2009) The neural basis of Drosophila gravity-sensing and hearing. Nature 458(7235):165–171. http://www.nature.com/nature/journal/v458/n7235/suppinfo/nature07810_S1.html

Keil TA (1997) Functional morphology of insect mechanoreceptors. Microsc Res Tech 39(6):506–531. 10.1002/(SICI)1097-0029(19971215)39:6<506::AID-JEMT5>3.0.CO;2-B

Kim SE, Coste B, Chadha A, Cook B, Patapoutian A (2012) The role of Drosophila piezo in mechanical nociception. Nature 483(7388):209–212. https://doi.org/10.1038/nature10801

Li J, Zhang W, Guo Z, Wu S, Jan LY, Jan YN (2016) A defensive kicking behavior in response to mechanical stimuli mediated by Drosophila wing margin bristles. J Neurosci 36(44):11275–11282. https://doi.org/10.1523/JNEUROSCI.1416-16.2016

Lumpkin EA, Marshall KL, Nelson AM (2010) Review series: the cell biology of touch. J Cell Biol 191(2):237–248. https://doi.org/10.1083/jcb.201006074

Menon C, Brodie R, Clift S, Vincent JFV (2009) Concept design of strain sensors inspired by campaniform sensilla. Acta Astronaut 64(2–3):176–182. https://doi.org/10.1016/j.actaastro.2008.07.007

Robertson JL, Tsubouchi A, Tracey WD (2013) Larval defense against attack from parasitoid wasps requires nociceptive neurons. PLoS One 8(10):e78704. https://doi.org/10.1371/journal.pone.0078704

Singhania A, Grueber WB (2014) Development of the embryonic and larval peripheral nervous system of Drosophila. Wiley Interdiscip Rev Dev Biol 3(3):193–210. https://doi.org/10.1002/wdev.135

Suslak TJ, Watson S, Thompson KJ, Shenton FC, Bewick GS, Armstrong JD et al (2015) Piezo is essential for amiloride-sensitive stretch-activated mechanotransduction in larval Drosophila dorsal bipolar dendritic sensory neurons. PLoS One 10(7):e0130969. https://doi.org/10.1371/journal.pone.0130969

Tsubouchi A, Caldwell Jason C, Tracey WD (2012) Dendritic filopodia, ripped pocket, NOMPC, and NMDARs contribute to the sense of touch in drosophila Larvae. Curr Biol 22(22):2124–2134. doi:http://dx.doi.org/10.1016/j.cub.2012.09.019

von Philipsborn AC, Liu T, Yu JY, Masser C, Bidaye SS, Dickson BJ (2011) Neuronal control of Drosophila courtship song. Neuron 69(3):509–522. https://doi.org/10.1016/j.neuron.2011.01.011

Xiang Y, Yuan Q, Vogt N, Looger LL, Jan LY, Jan YN (2010) Light-avoidance-mediating photoreceptors tile the Drosophila larval body wall. Nature 468(7326):921–926. https://doi.org/10.1038/nature09576

Yan Z, Zhang W, He Y, Gorczyca D, Xiang Y, Cheng LE et al (2013) Drosophila NOMPC is a mechanotransduction channel subunit for gentle-touch sensation. Nature 493(7431):221–225. https://doi.org/10.1038/nature11685

Yorozu S, Wong A, Fischer BJ, Dankert H, Kernan MJ, Kamikouchi A et al (2009) Distinct sensory representations of wind and near-field sound in the Drosophila brain. Nature 458(7235):201–205. http://www.nature.com/nature/journal/v458/n7235/suppinfo/nature07843_S1.html

Zhang W, Yan Z, Jan LY, Jan YN (2013) Sound response mediated by the TRP channels NOMPC, NANCHUNG, and INACTIVE in chordotonal organs of Drosophila larvae. Proc Natl Acad Sci U S A 110(33):13612–13617. https://doi.org/10.1073/pnas.1312477110

Zhang W, Yan Z, Li B, Jan LY, Jan YN (2014) Identification of motor neurons and a mechanosensitive sensory neuron in the defecation circuitry of Drosophila larvae. elife 3. https://doi.org/10.7554/eLife.03293

Chapter 4
Mechanotransduction in *Drosophila* Mechanoreceptors

Abstract In Chap. 3, we discussed the general physiology of fly mechanoreceptors. You may ask "well, these cells are sensitive to forces, but…how?" Great question! In Chap. 4, we will introduce the working mechanisms of several fly mechanoreceptors, including the bristle sensilla, campaniform sensilla, chordotonal organs, and class I–IV da neurons in fly larvae. We will cover the structural, mechanical, and molecular basis of mechanotransduction in these mechanoreceptors. On one hand, these mechanoreceptors share common structural designs, mechanical principles, and mechanotransduction molecules. On the other hand, they have also developed their own characters to fit the specific sensory functions. After reading this chapter, one might wonder about the stunning structure-mechanical design of the fly mechanoreceptors. Furthermore, these mechanoreceptors might share similar mechanics with the vertebrate or mammalian mechanoreceptors, so they provide great models to study the general principles in mechanotransduction.

4.1 Overview of Fly Mechanotransduction

We have discussed about the minimal mechanotransduction apparatus in Chap. 2. Despite being intuitive and theoretical, the conceptual apparatus is useful when we consider the structural, mechanical, and functional designs of the real mechanosensory cells by helping to sort out the different components (cells, structures, molecules, etc.) in a theoretical framework. If we try to understand fly mechanoreceptors as an ideal mechanosensor, we may ask: (1) What are the structures or molecules that form the mechanical processor? (2) What molecules act as the responders? (3) How does the processor mechanically work together with the responder? Taking these questions, let's go through the fly mechanoreceptors together.

© The Author(s) 2017

X. Liang et al., *Mechanosensory Transduction in Drosophila Melanogaster*,
SpringerBriefs in Biochemistry and Molecular Biology,
DOI 10.1007/978-981-10-6526-2_4

4.2   Bristle Receptor

Bristle sensilla respond to the deflection of the hair caused by the direct touch (Lumpkin et al. 2010; Gillespie and Walker 2001; Walker et al. 2000; Keil 1997). In the first step, the hair deflection is converted to a compressive force to the sensory cell at the base of the hair. Then, the sensory cell converts the force to a change in the receptor potential at the transduction site. If the stimulation is strong enough, the receptor potential will propagate and induce the neuronal action potential that finally excites the neuron and initiates the sensory input. We start our discussion on bristle mechanotransduction from the hair deflection.

4.2.1   Bristle Deflection

Bristles and hairs are the external cuticle structures of the bristle sensilla. They are constructed as the first-order levers (Box 4.1) (Keil 1997). In such a structural design, a large movement of the hair tip (distal end) is transformed into a small movement of the proximal base that directly compresses or stretches the dendritic tip of the sensory neuron. The key factors for a lever are the pivot point and two lever arms. In this view, we can discuss the structural and mechanical design of the bristle sensilla.

In bristle sensilla, the hair base has an oval shape (in the cross-sectional view) with the lateral protrusions that abut the socket-like cuticle structures (Fig. 4.1a)

Box 4.1: Three Types of Levers
First-order lever: the pivot is between the effort and the load. This is the simplest form of lever and the prototype of "lever" concept in our mind, for example, a crowbar.

Second-order lever: the load is between the effort and the pivot, for example, a wrench.

Third-order lever: the effort is between the load and the pivot, for example, a forceps.

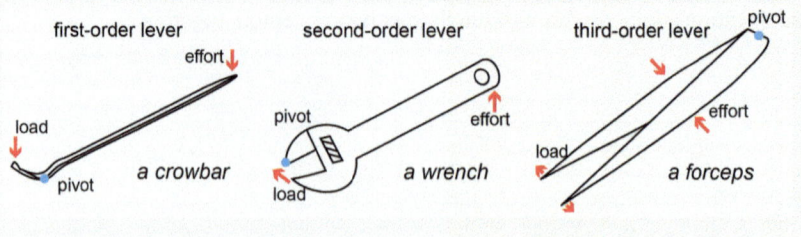

Fig. 4.1 Overview of the bristle mechanics. (**a**) Structural organization of the bristle base and the dendritic tip. (**b**) The hair movement is restricted only in the XZ plane, not in the XY and YZ planes. (**c**) The bristle is equivalent to a lever. A *black dot* indicates the pivot point, and two lever arms are indicated with the *blue lines*. (**d**) The mechanical schematic of a bristle lever (d_d distal deflection, f_d distal force, d_b basal deflection, f_b basal force)

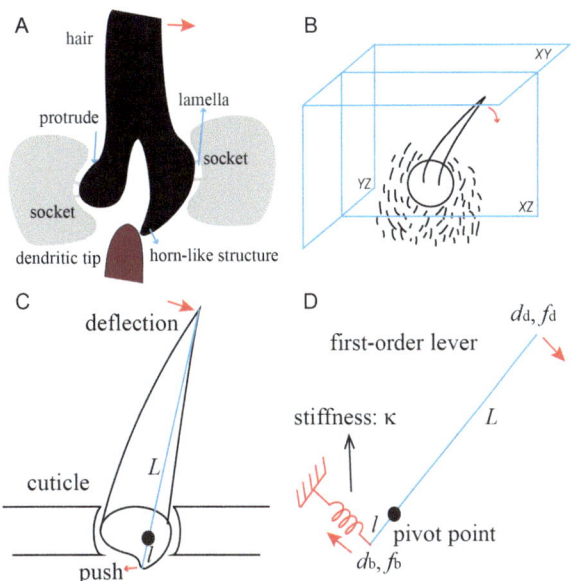

(Keil 1997). These two protrusions are linked together with the socket through a layer of lamella and some connection fibers, both together called the joint membrane (Fig. 4.1a). This protrusion-socket complex has three functions. First, in this way, the hair base is somehow fixed to the socket so their movements are constrained. Therefore, this spot can be considered as of forming a pivot point if we consider the hair as a lever. Second, it constraints the hair movement only in a specific plane (the XZ plane) that is perpendicular to the body surface (the XY plane) and to the axis of the protrusion-socket complex (the YZ plane in Fig. 4.1b). This structural polarity shapes the directional sensitivity of the sensory cell in the bristle sensillum (Walker et al. 2000; Keil 1997). Third, the joint membrane is elastic, so it mechanically contributes to the transformation of hair deflection to the basal compressive force, possibly by functioning as a cushion that damps the basal deformation. This can be a protection mechanism for the dendritic tip. It could also act as an elastic spring that helps to restore the hair back to the resting position after the stimuli. Therefore, the bristle has a defined pivot point and a lever plane.

At the basalmost of the hair, the hair base forms a narrow hornlike structure that makes a direct contact with the dendritic tip of the sensory cell (Fig. 4.1a, b). The axial distance between this structure and the "pivot" varies in different types of bristle sensilla, likely in the range of 1–2 μm (Keil 1997). This can be considered as the length of the short lever arm. The bristles themselves are the long lever arms. In *Drosophila* macrochaetes, the bristles are around 100–150 μm. There're also microchaetes with much shorter bristles. These bristles are rich in cytoskeleton bundles and thereby likely have a rigidity over GPa (Howard 2001), more rigid than the dendritic tip. Therefore, the bristle sensillum has a clear lever axis, a pivot point, and two rigid lever arms (Fig. 4.1c, d). The pivot point locates between the bristle tip and the hornlike structure, which makes the bristles mechanically a first-order lever with the lever ratio of 50–100 in *Drosophila* macrochaetes.

Given that the bristle can be considered as a first-order lever, we then obtain (Eq. 4.1, also see Fig. 4.1d)

$$\frac{d_b}{d_d} = \frac{l}{L} \tag{4.1}$$

Taking a 100 μm long bristle as an example, its hornlike structure is about 2 μm away from the pivot point, so the lever ratio is about 50. This means that when the bristle tip is deflected for 100 nm, the horn moves for 2 nm. This is a useful transformation because the dendritic tip, in the axis of the compressive force, has a width of only about 200–300 nm; it is impossible to take up 100 nm deformation without destroying the tip. In this case, the bristle matches the impedance and transforms the 100 nm deflection to the 2 nm deformation for the dendritic tip.

We also have (Eq. 4.2, also see Fig. 4.1d)

$$\frac{f_d}{f_b} = \frac{l}{L} \tag{4.2}$$

This means that the force increases 50 times, so that the distal signal is amplified by the bristle lever when it arrives at the transduction site. Furthermore, if we consider the compound stiffness of the dendritic tip (together with all supporting materials) is κ_b:

$$\kappa_b = \frac{f_b}{d_b} = \left(\frac{L}{l}\right)^2 \cdot \frac{f_d}{d_d} = \left(\frac{L}{l}\right)^2 \cdot \kappa_d \tag{4.3}$$

Because the lengths of two lever arms are measurable, if the force-deflection curve of the bristle tip can be measured, the equivalent stiffness of the dendritic tip can be estimated (Eq. 4.3). For example, if the equivalent stiffness of the bristle tip is 1 pN/nm and the lever ratio is 50, then the dendritic tip and its associated structures (including the joint membrane) would have a compound stiffness of 2.5 nN/nm. This is useful because by comparing the ultrastructure of wild-type and mutant flies, this mechanical measurement, together with the ultrastructural analysis, will lead the mechanic studies on the bristle sensilla to the molecular level (also see below in the campaniform sensilla part).

4.2.2 Dendritic Tip and the Supporting Structures

The dendritic tip of the sensory neuron in the bristle sensillum is likely the site of mechanotransduction. First, a strong piece of evidence is that the transduction channel in bristle sensilla, i.e., NompC, exclusively localizes at the dendritic tip (Liang et al. 2011; Lee et al. 2010). Second, this tip structure has a direct contact with the hornlike structure at the hair base and thereby at the right position to receive the compressive forces caused by the horn movement (Keil 1997). Third, this tip structure is surrounded by a set of supporting structures and has a delicate intracellular structure, which likely support its sensory functions (Liang et al. 2013; Keil 1997).

The cuticular structures that support the dendritic tip are quite complex (Keil 1997; Thurm et al. 1983). The dendritic tip is enclosed by a layer of dendritic sheath (Fig. 4.2a) secreted by one of the supporting cells, i.e., the thecogen cell. The hornlike structure indents the dendritic sheath and makes the direct contact with the dendritic tip (Fig. 4.2a). On the opposite side of the hornlike structure, there is a cap structure that likely reinforces the dendritic tip to the compressive forces (Fig. 4.2a). Surrounding the dendritic tip and sheath is the socket septum in the circular cuticular socket structures. The socket septum is made of collagen-like materials and is thought to function like a cushion that encloses the entire dendritic tip.

The cytoplasm of the dendritic tip has a bundle of microtubules (Fig. 4.2a). The spaces between the individual microtubules are filled with the electron-dense mate-

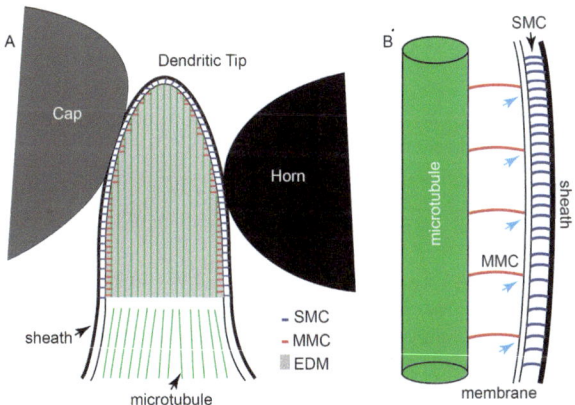

Fig. 4.2 The organization of the dendritic tip and its supporting structures. (**a**) The hornlike structure indents the dendritic tip. The cap structure localizes at the opposite site of the hornlike structure, likely to provide a mechanical support for the dendritic tip. The cytoplasm of the dendritic tip is filled with a highly ordered and microtubule-based structure. For extracellular side to intracellular side, it contains the dendritic sheath, sheath-membrane connector (SMC), membrane, membrane-microtubule connector (MMC), microtubule, and electron-dense materials. (**b**) The enlarged schematic for the microtubule-based structure. The extracellular forces transmit through the sheath, SMC, membrane, MMC, and microtubules. The MMCs focus the forces to the discrete spots (*blue arrowheads*) on the membrane where the transduction channel should localize

rial made of unknown substances. Most importantly, around the horn-tip contact site, there are intracellular membrane-microtubule connectors (MMCs) and the sheath-membrane connectors (SMCs) (Fig. 4.2a, b) (Keil 1997; Thurm et al. 1983). These connectors link all structural components in the dendritic tip into a structural entirety, which provides a structural basis for the mechanical signaling pathway. In addition, these connectors are only found in this region, suggesting their specific sensory roles.

When the hair is touched, the hair deflection is transformed by its lever-like structure into a small oppositely directed movement of the hair base. In this way, the hornlike structure compresses the dendritic tip through the SMCs, membrane, and the MMCs (Fig. 4.2b), which is thought to cause the depolarization of the dendritic membrane and eventually elicit the neuronal spikes. The first membrane depolymerization is thought to be caused by the opening of force-gated mechanosensitive ion channels that localize at the proximity of the MMCs. In the last few years, it was shown that NompC/TRPN is likely the mechanotransduction channel at the dendritic tip. More interestingly, the ankyrin-repeat domain of NompC structurally contributes to the MMCs (Liang et al. 2011, 2013; Lee et al. 2010; Walker et al. 2000), thereby suggesting a force transmission pathway to the NompC channels. Because most of the ultrastructural studies were performed on the campaniform sensilla and they have similar structural features as the bristle sensilla, we will elaborate on the ultrastructure in the following discussions of campaniform sensilla.

4.3   Campaniform Receptor

Campaniform sensilla respond to the cuticle deformation, more precisely the curvature change of the cuticle surface caused by the relative movements of the body segments during walking, jumping, and flying (Fayyazuddin and Dickinson 1996). The cuticle deformation is transformed into a compressive or stretching force to the sensory cells in the campaniform sensillum (Keil 1997). Similar to the bristle sensilla, the sensory cells read the mechanical signal and output the neuronal impulses to initiate the sensory feedback inputs. Campaniform receptors are great models to study the ultrastructure of fly mechanoreceptors because they often enrich in receptor fields (Liang et al. 2013; Keil 1997), which makes the sample preparation for electron microscopy more straightforward. We will summarize our current knowledge of the campaniform sensilla.

4.3.1   Cuticle Deformation

Campaniform sensilla at different parts of the fly body have different cuticle morphologies (see Chap. 3, Fig. 4.2), but they function in a similar way. We take the campaniform sensilla in the haltere pedicellum as the examples (Liang et al. 2013).

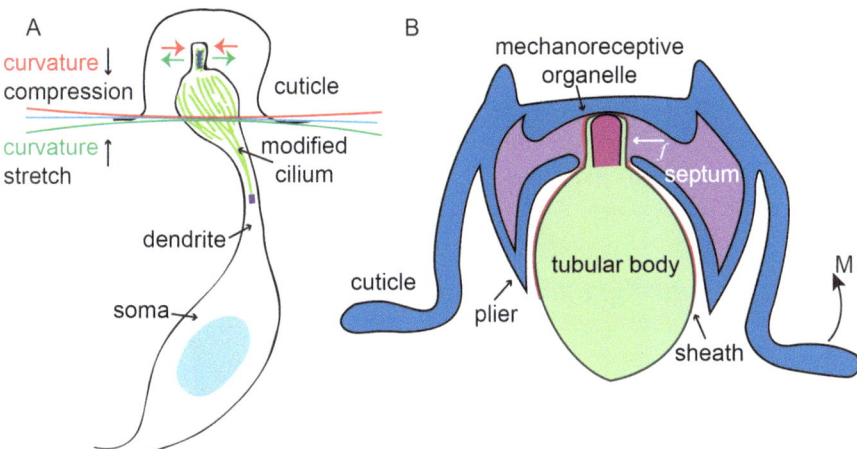

Fig. 4.3 The cuticle deformation is transformed into the compressive or stretching forces to the dendritic tip. (**a**) The increase in the cuticle curvature (*green*) creates a stretching force (*green arrows*) to the mechanoreceptive organelle. The decrease in the cuticle curvature (*red*) creates compressive forces (*red arrows*) to the mechanoreceptive organelle. (**b**) The cuticle structure (*blue*) and the socket septum (*purple*) together transform the torque (*M*, caused by the cuticle deformation) stimuli into forces onto the dendritic tip (only the stretching scenario is shown in this schematic diagram)

In the haltere pedicellum, there are several receptor fields that consist of many campaniform receptors (see Chap. 3, Fig. 4.2). The longitudinal sectional view of the campaniform sensillum shows a Ω-shaped cuticle structure whose arms extend toward both sides and fuse to the Ω-shaped cuticle structures of the adjacent sensilla (Fig. 4.3). The dendritic tip of the sensory cell abuts the cuticle structure (Fig. 4.3a). Like in the bristle sensilla, there are the socket septum materials between the cuticle and dendritic sheath that encloses the dendritic tip (Fig. 4.3b). The socket septum here may have similar roles as the septum in the bristle and may also share the functions of the joint membrane that damps the incoming stimuli as a protection mechanism. Through the cuticle, socket septum, and dendritic sheath, the cuticle deformation is transformed to the forces onto the dendritic tip of the sensory cell (Fig. 4.3a, b).

Because the Ω-shaped cuticle structures of the pedicel campaniform mechanoreceptor stand out from the curved cuticle surface of the haltere, the increase in the cuticle curvature (bend downward, like the green line in Fig. 4.3a) tends to stretch the upper surface of the cuticle and thereby creates a stretching force to the dendritic tip, while the decrease in the cuticle curvature (bend upward, like the red line in Fig. 4.3a) has the opposite effects (i.e., compressive force to the dendritic tip) (Fig. 4.3a). The Ω-shaped cuticle structure basically acts as a second-order lever (Box 4.1) that transforms the cuticle deformation into the stretching/compressive forces onto the cell membrane of the mechanoreceptive organelle. Intuitively, when the cuticle structure receives a stretching force, it acts in a similar way as a bottle opener in which the dendritic tip membrane is like the "lid" to open. The pivot point of this "lever" is provided at the top-center region of the cuticle (Fig. 4.3b). The Ω-shaped cuticle has two symmetrically arranged arched collars, each of which functions as the longer arm of the lever-like structure. When the cuticle is stretched or compressed, the Ω-shaped lever arm receives a torque (M, due to the cuticle stretch or compression) and outputs the force onto the dendritic tip (Fig. 4.3b).

4.3.2   Dendritic Tip and the Supporting Structures

Similar to the bristle receptors, the dendritic tip of the sensory neuron is likely the site of mechanotransduction as it abuts the cuticle structure and is ready to receive the forces caused by the cuticle deformation. Around the turning point of the Ω-shaped cuticle (Fig. 4.3b), there are a pair of plier structures made of similar materials as the cuticle. This pair of pliers extends inward to the dendritic tip (Fig. 4.3b). The pliers do not have a direct contact with the dendrite but connect to it through the socket septum (Fig. 4.3b). This further suggests that the septum may contribute to the overall working mechanics of these sensilla by damping the deformation to some extent and thereby acting as a protection mechanism to the sensory organelle.

The dendritic tip of campaniform receptor is among the best studied mechanoreceptive organelles (Liang et al. 2013). The entire dendritic tip is a modified cilium that contains four major parts, the cilium, tubular body, neck, and mechanoreceptive organelle (Fig. 4.4a). These four segments contain the microtubule-based structures with different spatial organizations (Fig. 4.4a). The key structural features of these segments include:

1. The cilium contains a "9 + 0" axonemal structure that is typical for the sensory cilium in fly. The ciliary structure separates the dendrite into the inner (basal to the cilium) and the outer (distal to the cilium) segments (Fig. 4.4a).
2. On the distal side of the cilium, there exists a dilated structure which contains hundreds of microtubules. These microtubules are linked by the filamentous linkers and are organized in a highly ordered manner. This dilated structure is named as the "tubular body." The exact function of this organelle is not clear so far, but we suspect that it may act as a rigid base that mechanically supports the mechanoreceptive organelle. It is also not clear how this organelle is generated, for example, how these hundreds of microtubules are nucleated, polymerized, stabilized, and ordered (Fig. 4.4a).
3. The neck structure contains relatively irregular microtubules whose orientations are not as regular as those in the tubular body and in the mechanoreceptive organelle. This segment appears to be a transition structure between the tubular body and the mechanoreceptive organelle, so it may not be directly involved in the mechanotransduction (Fig. 4.4a).

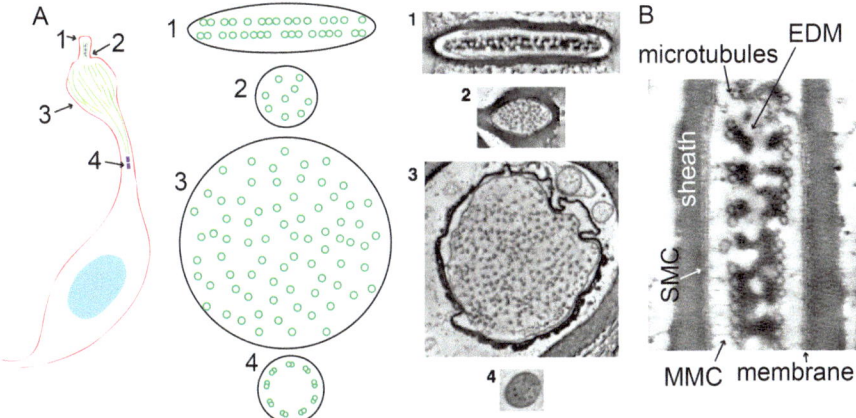

Fig. 4.4 The ultrastructure of the modified cilium at the distal end of the dendrite in the campaniform receptor. (**a**) The modified cilium consists of four major segments: (*1*) the mechanoreceptive organelle, (*2*) the neck structure, (*3*) the tubular body, and (*4*) the cilium. Each of these segments contains a microtubule-based cytoskeleton structure. The *left panel* shows the location of the modified cilium; the *middle panel* shows the cross-sectional view of the microtubule-based structure in which each green circle represents a microtubule; the *right panel* shows the TEM micrographs for regions 1–4 in which the microtubules are clearly visible. (**b**) The fine structures in the mechanoreceptive organelle (*MMC* membrane-microtubule connector, *SMC* sheath-membrane connector, *EDM* electron-dense material)

4. The mechanoreceptive organelle has a nearly oval shape (cross-sectional view) with a longer axis and a shorter one. The asymmetry in the shape of the mechanoreceptive organelle provides a structural basis for the directional sensitivity of the campaniform sensilla, namely, this type of sensilla primarily responds to the stimuli perpendicular to its longer axis. Furthermore, the mechanoreceptive organelle contains a delicate intracellular structure. The main intracellular components are two paralleled rows of microtubules. Between two rows of microtubules are electron-dense materials made of unknown substances. The microtubules are organized into small arrays (in average 4–5 microtubules) through the tiny linkers (each about 3 nm long) between the adjacent microtubules. The microtubules are linked to the cell membrane through the filamentous connectors called the "membrane-microtubule connectors" (MMCs). The membrane is connected to the dendritic sheath (extracellular) through the small "sheath-membrane connectors" (SMCs). MMCs and SMCs together organize the mechanoreceptive organelle into a structural entirety, which is important for the entire organelle to function as a force sensor (Fig. 4.4a, b).

4.3.3   Molecular Basis of Mechanotransduction

At the molecular level, the mechanistic nature of mechanotransduction in campaniform receptor resides on two questions: what is the transduction channel and how is the transduction channel gated by force?

So far, all the experimental evidences consistently suggest that NompC/TRPN is likely the mechanotransduction channel in the bristle and campaniform receptors. The evidence include (1) the mechanotransduction in the bristle and campaniform mechanoreceptors in *nompC* null mutants is nearly abolished (90% reduction in bristle receptors and nearly 100% reduction in campaniform receptors) (Liang et al. 2013; Walker et al. 2000); (2) a missense point mutant in the *nompC* gene causes a faster adaptation of the bristle receptors (Walker et al. 2000); (3) putting the *nompC* gene or cDNA back to the genome rescues the null mutant phenotypes (Liang et al. 2013; Cheng et al. 2010); (4) NompC exclusively localizes at the mechanoreceptive organelle that has a direct contact with the cuticle structures (Liang et al. 2011, 2013; Lee et al. 2010; Cheng et al. 2010); and.(5) NompC, by itself, is a bona fide mechanosensitive ion channel (Zhang et al. 2014; Yan et al. 2013). Therefore, NompC is well supported to be the primary mechanotransduction channel in these two types of sensilla. One remaining mystery is what contributes to the 10% responsive current remained in the bristle receptors of the *nompC* null mutants (Walker et al. 2000). There is so far no evidence to support that there is a second type of mechanotransduction channel in the bristle or campaniform mechanoreceptors.

The N-terminal of NompC has a large ankyrin-repeat domain that consists of 29 ankyrin repeats (Jin et al. 2017; Howard and Bechstedt 2004; Walker et al. 2000). Because this ankyrin repeat domain was predicted to have a helical shape (Jin et al. 2017; Howard and Bechstedt 2004), it is usually termed as the ankyrin helix. It was later found that the ankyrin helix structurally contributes to the MMCs observed in the mechanoreceptive organelle of campaniform mechanoreceptors (Liang et al. 2013). As we just mentioned, MMCs focus the forces onto discrete spots on the membrane. This indicates that NompC has a direct connection to the MMCs and localizes at the right place where the membrane receives the largest forces. However, because these spots are also the place where the membrane is strained the most, it is still difficult to distinguish how NompC is gated (by pushing-pulling force or lateral membrane tension). More detailed analysis and discussion on the gating mechanism of NompC will be provided in Chap. 5.

Microtubules are the major structural components in the mechanoreceptive organelle. In fact, it also has a sensory function by mechanically reinforcing the MMCs so that the MMCs, possibly being compliant, are able to transmit the reflective forces back to the transduction channels in the membrane (Liang et al. 2014). In a mutant strain of the *dcx-emap* gene, 50% of the microtubules in the distal tip are lost, and the electron-dense materials are absent. Despite these defects, NompC can still localize to the right place, and the remaining microtubules have normal MMCs. Because there are still about 50% NompC-microtubule complexes present in the mechanoreceptive organelle, it was expected that the mutant campaniform mechanoreceptors have an intermediate phenotype in mechanotransduction. However, the mechanosensory response is reduced for more than 90% in the *dcx-emap* mutant strain. One possible explanation is that the microtubules, though present in the mutant mechanoreceptive organelle, lack the mechanical reinforcement from the electron-dense material, so they might contribute additionally to the total compliance of the sensory system and thereby lead to the reduction in the forces transmitted to the membrane. Therefore, the observations in the *dcx-emap* mutant suggest that the microtubules are not only structural components but also have sensory roles.

Based on the current information, a highly simplified and hypothetic model ("linkage model") is proposed for the mechanotransduction in campaniform mechanoreceptor (Liang et al. 2013, 2014), which is possibly also applicable for the bristle receptors. In this model, the distal mechanical signal in the environment is transformed by a lever-like structure to a proximal stimulus that acts on the dendritic tip of the sensory neuron (Figs. 4.1, 4.3). At the dendritic tip, the extracellular forces are transmitted through a set of linked structural components and finally arrive at the NompC transduction channels via the MMCs (Fig. 4.5). In this mechanical schematic, either the compressive forces or the lateral tension or both may contribute to the proximal stimuli for the transduction channels.

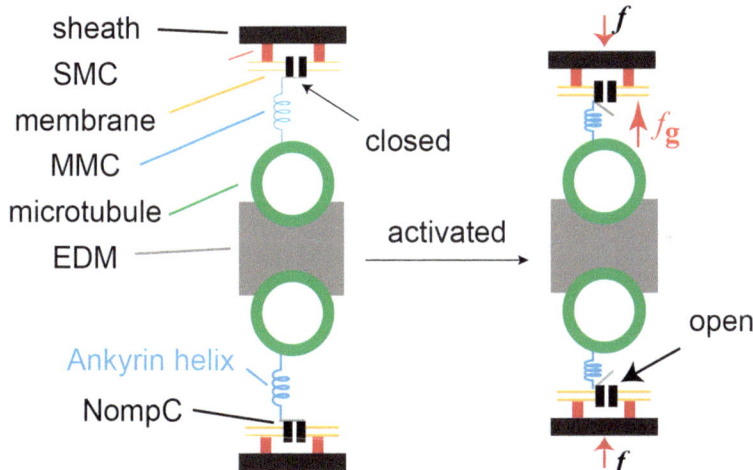

Fig. 4.5 The molecular diagram of the "linkage model." The extracellular force is transformed to the gating force by this molecular transduction apparatus onto the transduction channels. The ankyrin helix might function as a "gating spring" that takes up the deformation and regulates the gating force (*red arrow* in the *right panel*)

4.4   Chordotonal Organ

The chordotonal organ (also called the scolopidial organ) is the stress sensor that often localizes at the joints and reports the relative movements of the body segments (Keil 1997). However, the biggest chordotonal organ in fly is Johnston's organ, fly's auditory organ. It localizes in the second segment of the antenna (Fig. 4.6a) and is named after the physician Christopher Johnston. Johnston's organ is a great model system to study the molecular mechanics of the auditory system because of several

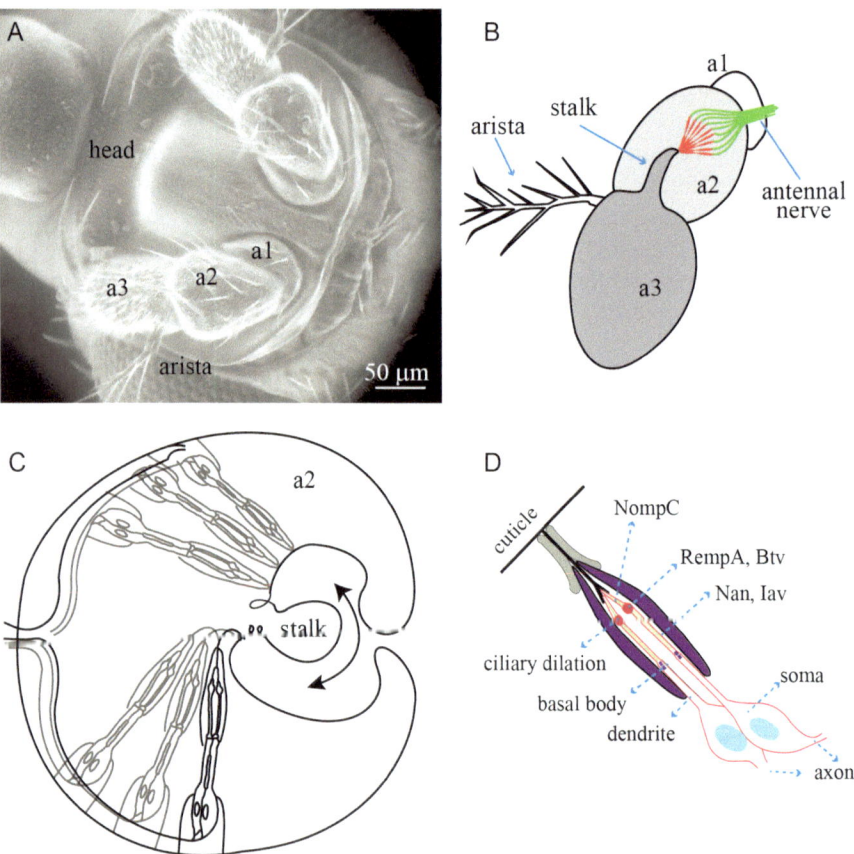

Fig. 4.6 Johnston's organ. (**a**) A SEM micrograph of two antennas on a fly head. The segments and the arista are indicated in the picture. (**b**) The structural organization of the antenna. Note that the arista is associated with the third antennal segment and the stalk inserts into the second segment. The distal ends of all scolopidial units physically connect to the stalk structure. (**c**) The cross-sectional view of the second antennal segments. Only three sensory units are shown here to represent one of the two populations that connect to the anterior or posterior side of the stalk. The rotation of the stalk will stretch one group of sensory cells and compress the other group. (**d**) The functional anatomy of the scolopidial units. The key structures are indicated on the left side, while the localizations of some key molecules are labeled on the right side

reasons: (1) the successful application of a noninvasive method for recording the auditory mechanics on fly antenna by using the laser Doppler interferometry (Nadrowski et al. 2008; Albert et al. 2007; Gopfert et al. 2006), (2) the *Drosophila* genetic tools, and (3) the similarities and differences between Johnston's organ and vertebrate hair cells in hearing mechanics (Bokolia and Mishra 2015; Bechstedt and Howard 2008). Apart from being the auditory organ, Johnston's organ is also responsible for the wind and gravity sensation (Yorozu et al. 2009; Kamikouchi et al. 2009). We take the Johnston's organ as a typical example to introduce the chordotonal organs.

4.4.1 Fly Antenna and Johnston's Organ

In *Drosophila*, the third antennal segment (a3 in Fig. 4.6a, b) and its associated arista serve as the sound receiver (Fig. 4.6a, b) (Bokolia and Mishra 2015; Nadrowski et al. 2008). In the presence of the acoustical stimuli, the vibrations of the arista drive the sympathetic rotation of the third segment back and forth around the joint between the second (a2 in Fig. 4.6a, b) and the third antennal segments. Because the third antennal segment and the arista can be nearly considered as a rigid entirety, their vibrations are coupled to Johnston's organ in the antennal second segment (Fig. 4.6b) and then transduced to the neuronal responses. The coupling is achieved through a stalk structure of the third segment that inserts into the second segment and has a physical connection with the sensory cells in Johnston's organ (Fig. 4.6b) (Boekhoff-Falk and Eberl 2014; Eberl and Boekhoff-Falk 2007). The scolopidial units are grouped into two populations that connect to the anterior and posterior sides of the stalk structure (Fig. 4.6c). When the stalk structure rotates in a certain direction, one population of cells is stretched while the other is compressed. This tissue organization provides a structural basis for a mechanical amplification mechanism. For example, when the stalk rotates to activate/inactivate the anterior/posterior sensory cells, respectively, these activated/inactivated cells provide the mechanotransduction-related reduction/increase in their stiffness, which would facilitate the stalk rotation and in turn the vibrations of the third segment and the arista (Nadrowski et al. 2008). In this way, the antenna is able to amplify the weak acoustic signals. This is in principle similar to the amplification mechanism in the vertebrate auditory organs (i.e., eardrum and ear bones), though the anatomies of the organs are vastly different (Bokolia and Mishra 2015).

4.4.2 Molecular Basis of Mechanotransduction

The structural organization of the dendritic tip (i.e., the modified cilium) has been introduced in Chap. 3; here, we review the functional anatomy of the modified cilium in the scolopidial unit (Fig. 4.6d). As described in Chap. 3, the modified cilium in the scolopidial cell contains a ciliary dilation that separates the cilium into the distal and proximal segments.

The distal segment has a direct connection to the stalk structure via the cap cell and the attachment cell (Chap. 3, Fig. 4.3). In the mutant strain of *NompA*, the connection between the distal segment and the cap cell is disrupted. This structural phenotype is accompanied by the complete loss of the mechanosensory responses to the acoustic stimuli, indicating that this physical connection is functionally required (Chung et al. 2001). One possibility is that the distal segment of the cilium is the site of mechanotransduction. This is additionally supported by the finding that NompC, a mechanotransduction channel, exclusively localizes at this subcellular region in the scolopidial sensory cells (Liang et al. 2011; Lee et al. 2010; Cheng et al. 2010). NompC expresses in nearly all sensory cells in Johnston's organ and is essential for the response of the sound receptors (Effertz et al. 2011). The mechanical measurements suggest that NompC is required for the mechanical feedback that ensures the exquisite sensitivity of the fly antenna (Gopfert et al. 2006) and is involved in the direct gating mechanism of the mechanotransduction apparatus at the molecular level (Effertz et al. 2012). Note that there is also a study that argues that NompC is not the transduction channel in the case of fly auditory cells (Lehnert et al. 2013). Interestingly, no MMC-like structures have been observed so far in this region. Therefore, further studies on the ultrastructural details on this region with the electron microscopy are certainly required for the integrated understanding of the structural mechanics.

The proximal side of the distal segment connects to the ciliary dilation (Fig. 4.6d). It is not definitely clear whether this subcellular structure has a mechanosensory function. One proposed mechanism is that when the cilium is stretched by the stalk movement, the transduction channels, if localized here, that are linked to the ciliary microtubules through the membrane-microtubule linkers are pushed or pulled against the paracrystalline materials inside the ciliary dilation. The compression or stretch would result in a force onto the membrane and activate the transduction channels to initiate the neuronal responses (Bechstedt et al. 2010). However, this model assumes the presence of mechanotransduction channels in the ciliary dilation, but only a small amount of NompC is observed in this region, and no other known channels localize here. Therefore, the ciliary dilation may not directly participate in the mechanotransduction but serve more structural or developmental roles in establishing or maintaining the polarity of the sensory cilium. In the mutant strain of *RempA*, the ciliary dilation is absent. *RempA* mutant strains lose the sound-evoked potentials in the antenna recording experiments (Lee et al. 2008). However, because *RempA* encodes an IFT protein (IFT140) and may have more general effects on the development of the cilium (e.g., the cilia are shorter in *RempA* mutants), so the observations on the *RempA* mutants do not specify the roles of the ciliary dilation in mechanotransduction. In the mutant strain of *btv*, the paracrystalline structure in the ciliary dilation is absent, and the *btv* mutants show severe defects in the sound-evoked potential recordings (Eberl et al. 2000). Despite these observations, the specific roles of the ciliary dilation in the scolopidial cell remain further elusive.

Future works on the ultrastructure of ciliary dilation and the mechanical model of the scolopidial sensory cilium will help in clarifying this issue.

The proximal segment of the cilium contains a typical axonemal structure and has important roles in mechanotransduction. Two important molecules, inactive (Iav) and nanchung (Nan), localize at the proximal segment (Fig. 4.6d) (Gong et al. 2004). Their mutants show the complete loss of the mechanosensory responses. Both inactive and nanchung belong to the TRP channel superfamily and are similar to the TRPV channels. These two TRP channels are interaction partners. This interaction is required for their normal functions and localizations. The mechanical measurements suggest that the Iav-Nan complex controls the power gain in the mechanical amplification and may serve in propagating the electrical signals to the nerve (Gopfert et al. 2006). In addition, this controlling function is dependent on the presence of NompC. However, the molecular and ultrastructural bases underlying the effects of Iav and Nan are not clear. It is also not clear how they work together with NompC to specify the sensitivity of Johnston's organ.

4.5 Dendritic Arborization Neurons

4.5.1 Overall Mechanics of da Neurons

All four types of dendritic arborization (da) neurons are mechanosensors (see Chap. 3, Table 3.1) (Yan et al. 2013; Tsubouchi et al. 2012; Kim et al. 2012; Cheng et al. 2010), among which the class III and class IV are best studied. The mechanics involved in their sensory processes is more straightforward due to their simpler tissue organizations. The dendrites of the da neurons are embedded in the epidermal layer with cuticle on one side and the extracellular matrix (ECM) on the other side (Fig. 4.7a, b) (Han et al. 2012). In some cases, the dendrites have a direct contact with the ECM, while in other cases, they are slightly away from the ECM. Class III and class IV neurons are touch sensors, so the distal stimuli for these cells are direct compressive deformations/forces (i.e., touch) onto the cuticle. Based on the contact mechanics, the proximal stimuli on the dendrites (i.e., pressure) are determined by the mechanical properties of the supporting tissues, including the cuticle, epidermal cell layer, and ECM (Fig. 4.7c). It is also important to consider the geometry of the contact interface as it determines the spatial distribution of the pressure in the stimulating field (Fig. 4.7c). In an ideal situation in which a sphere-shaped object compresses the cuticle, we obtain Eqs. 4.4 and 4.5, in which $p(r)$ is the spatial distribution of the pressure, F is the total force, and E is Young's modulus; other parameters are shown in Fig. 4.7c. Note that because the dendrites are embedded in the layers of the cuticle (~10 μm, $E = 0.4$ MPa), epidermis (~1 μm, $E = 10$ kPa), ECM (~0.1 μm, $E = 2$ GPa), and muscles (>100 μm, $E = 50$ kPa), Young's modulus E in Eq. 4.5 is a compound parameter that summarizes the mechanical properties of all these tissues. In a simple model, the forces or deformations caused by the compression of the

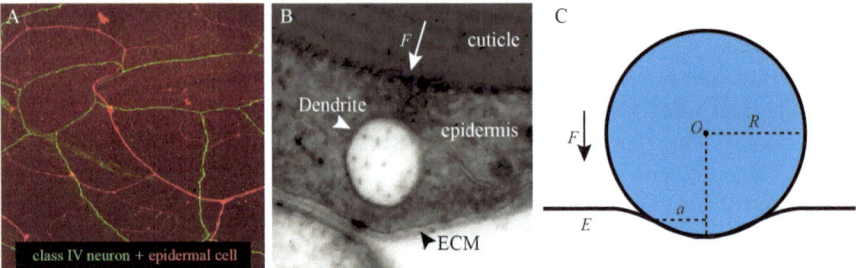

Fig. 4.7 The tissue organization of the dendritic arborization (da) neurons. (**a**) The class IV da neuron (*red*) is embedded in the epidermis (*green*). (**b**) A TEM micrograph shows the cross-sectional view of the dendrite of a class IV neuron with all structural components labeled. (**c**) A simple schematic for the contact mechanics that may be involved in the mechanosensory process of da neurons

cuticle surface are transmitted through the compound tissue to the dendrites. The mechanosensitive channels localizing at the dendritic membrane are then activated to initiate the neuronal responses. It should be noted that the tension applied to these structures by the underlying muscle fibers should, in principle, offset the sensitivity and the threshold of the sensory responses (Eq. 4.4, also see Fig. 4.7c and Eq. 4.5, also see Fig. 4.7c).

$$p(r) = p_0 \cdot \left(1 - \frac{r^2}{a^2}\right)^{\frac{1}{2}} \tag{4.4}$$

$$p_0 = \frac{1}{\pi}\left(\frac{6 \cdot F \cdot E^2}{R^2}\right)^{\frac{1}{3}} \tag{4.5}$$

4.5.2   Molecular Basis of Mechanotransduction

NompC is an important molecular player in the mechanotransduction process of the da neurons. It is expressed in class I neuron and is required in locomotion behavior of fly larvae (Cheng et al. 2010). In class I neurons, the NompC signal was found to be forming a filamentous shape, suggesting it may follow the microtubules in the dendrites. This is consistent with the observations in the campaniform receptors. It also expresses in class III neuron and is required in the sensation of "gentle touches" (Yan et al. 2013). In class III neurons, the spikelike neurites are rich in NompC and thereby could be the mechanotransduction sites. Interestingly, the spikes are rich in F-actin (Tsubouchi et al. 2012), while NompC, according to the studies in campaniform receptors, usually couples with the microtubules. One possibility is that there're also small numbers of microtubules in the spikes, which still have to be experimentally confirmed. The function of the F-actin could be to support the fine spikelike neurites and also may contribute in recruiting the microtubules to these spikes. In addition, individual F-actin fiber is more compliant than microtubule (Howard 2001), which suggests a possibility that the F-actin-rich neurite may generally be softer that fits their function to sense the gentle touches. However, F-actin could also form bundles to increase their mechanical rigidity, so the ultrastructural studies on these spikelike mechanoreceptive organelles are required for better understanding the molecular mechanism of the gentle touch transduction.

NompC doesn't express in the class IV da neurons, but the ectopic NompC expression in class IV neuron confers the gentle touch mechanosensitivity to class IV neurons (Yan et al. 2013). In class IV neurons, several channel proteins have been identified important in the sensory functions, including DmPiezo, Pickpocket1 (Ppk), and Pickpocket26 (Ppk26) (Mauthner et al. 2014; Guo et al. 2014; Gorczyca et al. 2014; Kim et al. 2012), among which DmPiezo has been shown to be a pore-forming mechanosensitive channel (Kim et al. 2012; Coste et al. 2012). It's been proposed that DmPiezo mediates one pathway in mechanosensation, while the PpK and PpK26 function together in a paralleled pathway (Guo et al. 2014). It is not yet clear how these channels are gated in the class IV da neurons. The open questions include (1) why do the class IV da neurons need two paralleled pathways for mechanosensory transduction?, (2) are PpK and PpK26 able to form a mechanosensitive ion channels?, (3) what are the proximal stimuli for DmPiezo or PpK/PpK26, and do they also need to function in a specialized intracellular structure, just like in the case of NompC?

References

Albert JT, Nadrowski B, Göpfert MC (2007) Mechanical signatures of transducer gating in the Drosophila ear. Curr Biol 17(11):1000–1006. http://dx.doi.org/10.1016/j.cub.2007.05.004

Bechstedt S, Howard J (2008) Hearing mechanics: a fly in your ear. Curr Biol 18(18):R869–R870. https://doi.org/10.1016/j.cub.2008.07.069

Bechstedt S, Albert JT, Kreil DP, Muller-Reichert T, Gopfert MC, Howard J (2010) A doublecortin containing microtubule-associated protein is implicated in mechanotransduction in Drosophila sensory cilia. Nat Commun 1:11. https://doi.org/10.1038/ncomms1007

Boekhoff-Falk G, Eberl DF (2014) The Drosophila auditory system. Wiley Interdiscip Rev Dev Biol 3(2):179–191. https://doi.org/10.1002/wdev.128

Bokolia NP, Mishra M (2015) Hearing molecules, mechanism and transportation: modeled in Drosophila melanogaster. Dev Neurobiol 75(2):109–130. https://doi.org/10.1002/dneu.22221

Cheng LE, Song W, Looger LL, Jan LY, Jan YN (2010) The role of the TRP channel NompC in Drosophila larval and adult locomotion. Neuron 67(3):373–380. https://doi.org/10.1016/j.neuron.2010.07.004

Chung YD, Zhu J, Han Y, Kernan MJ (2001) nompA encodes a PNS-specific, ZP domain protein required to connect mechanosensory dendrites to sensory structures. Neuron 29(2):415–428

Coste B, Xiao B, Santos JS, Syeda R, Grandl J, Spencer KS et al (2012) Piezo proteins are pore-forming subunits of mechanically activated channels. Nature 483(7388):176–181. https://doi.org/10.1038/nature10812

Eberl DF, Boekhoff-Falk G (2007) Development of Johnston's organ in Drosophila. Int J Dev Biol 51(6–7):679–687. https://doi.org/10.1387/ijdb.072364de

Eberl DF, Hardy RW, Kernan MJ (2000) Genetically similar transduction mechanisms for touch and hearing in Drosophila. J Neurosci 20(16):5981–5988

Effertz T, Wiek R, Göpfert MC (2011) NompC TRP Channel is essential for Drosophila sound receptor function. Curr Biol 21(7):592–597. http://dx.doi.org/10.1016/j.cub.2011.02.048

Effertz T, Nadrowski B, Piepenbrock D, Albert JT, Gopfert MC (2012) Direct gating and mechanical integrity of Drosophila auditory transducers require TRPN1. Nat Neurosci 15(9):1198–1200. http://www.nature.com/neuro/journal/v15/n9/abs/nn.3175.html#supplementary-information

Fayyazuddin A, Dickinson MH (1996) Haltere afferents provide direct, electrotonic input to a steering motor neuron in the blowfly. Calliphora J Neurosci 16(16):5225–5232

Gillespie PG, Walker RG (2001) Molecular basis of mechanosensory transduction. Nature 413(6852):194–202. https://doi.org/10.1038/35093011

Gong Z, Son W, Chung YD, Kim J, Shin DW, McClung CA et al (2004) Two interdependent TRPV channel subunits, inactive and Nanchung, mediate hearing in Drosophila. J Neurosci 24(41):9059–9066. https://doi.org/10.1523/JNEUROSCI.1645-04.2004

Gopfert MC, Albert JT, Nadrowski B, Kamikouchi A (2006) Specification of auditory sensitivity by Drosophila TRP channels. Nat Neurosci 9(8):999–1000. https://doi.org/10.1038/nn1735

Gorczyca David A, Younger S, Meltzer S, Kim Sung E, Cheng L, Song W et al (2014) Identification of Ppk26, a DEG/ENaC channel functioning with Ppk1 in a mutually dependent manner to guide locomotion behavior in Drosophila. Cell Rep 9(4):1446–1458. http://dx.doi.org/10.1016/j.celrep.2014.10.034

Guo Y, Wang Y, Wang Q, Wang Z (2014) The role of PPK26 in Drosophila larval mechanical nociception. Cell Rep 9(4):1183–1190. http://dx.doi.org/10.1016/j.celrep.2014.10.020

Han C, Wang D, Soba P, Zhu S, Lin X, Jan LY et al (2012) Integrins regulate repulsion-mediated dendritic patterning of drosophila sensory neurons by restricting dendrites in a 2D space. Neuron 73(1):64–78. https://doi.org/10.1016/j.neuron.2011.10.036

Howard J (2001) Mechanics of motor proteins and the cytoskeleton. Sinauer Associates, Publishers, Sunderland

Howard J, Bechstedt S (2004) Hypothesis: a helix of ankyrin repeats of the NOMPC-TRP ion channel is the gating spring of mechanoreceptors. Curr Biol 14(6):R224–R226. https://doi.org/10.1016/j.cub.2004.02.050

Jin P, Bulkley D, Guo Y, Zhang W, Guo Z, Huynh W et al (2017) Electron cryo-microscopy structure of the mechanotransduction channel NOMPC. Nature. https://doi.org/10.1038/nature22981

Kamikouchi A, Inagaki HK, Effertz T, Hendrich O, Fiala A, Gopfert MC et al (2009) The neural basis of Drosophila gravity-sensing and hearing. Nature 458(7235):165–171. http://www.nature.com/nature/journal/v458/n7235/suppinfo/nature07810_S1.html

Keil TA (1997) Functional morphology of insect mechanoreceptors. Microsc Res Tech 39(6):506–531. https://doi.org/10.1002/(SICI)1097-0029(19971215)39:6<506::AID-JEMT5>3.0.CO;2-B

Kim SE, Coste B, Chadha A, Cook B, Patapoutian A (2012) The role of Drosophila Piezo in mechanical nociception. Nature 483(7388):209–212. https://doi.org/10.1038/nature10801

Lee E, Sivan-Loukianova E, Eberl DF (2008) Kernan MJ. An IFT-A protein is required to delimit functionally distinct zones in mechanosensory cilia. Curr Biol 18(24):1899–1906. http://dx.doi.org/10.1016/j.cub.2008.11.020

Lee J, Moon S, Cha Y, Chung YD (2010) Drosophila TRPN(=NOMPC) channel localizes to the distal end of mechanosensory cilia. PLoS One 5(6):e11012. https://doi.org/10.1371/journal.pone.0011012

Lehnert BP, Baker AE, Gaudry Q, Chiang AS, Wilson RI (2013) Distinct roles of TRP channels in auditory transduction and amplification in Drosophila. Neuron 77(1):115–128. https://doi.org/10.1016/j.neuron.2012.11.030

Liang X, Madrid J, Saleh HS, Howard J (2011) NOMPC, a member of the TRP channel family, localizes to the tubular body and distal cilium of Drosophila campaniform and chordotonal receptor cells. Cytoskeleton (Hoboken) 68(1):1–7. https://doi.org/10.1002/cm.20493

Liang X, Madrid J, Gartner R, Verbavatz JM, Schiklenk C, Wilsch-Brauninger M et al (2013) A NOMPC-dependent membrane-microtubule connector is a candidate for the gating spring in fly mechanoreceptors. Curr Biol 23(9):755–763. https://doi.org/10.1016/j.cub.2013.03.065

Liang X, Madrid J, Howard J (2014) The microtubule-based cytoskeleton is a component of a mechanical signaling pathway in fly campaniform receptors. Biophys J 107(12):2767–2774. https://doi.org/10.1016/j.bpj.2014.10.052

Lumpkin EA, Marshall KL, Nelson AM (2010) Review series: the cell biology of touch. J Cell Biol 191(2):237–248. https://doi.org/10.1083/jcb.201006074

Mauthner Stephanie E, Hwang Richard Y, Lewis Amanda H, Xiao Q, Tsubouchi A, Wang Y et al (2014) Balboa binds to pickpocket in vivo and is required for mechanical nociception in Drosophila larvae. Curr Biol 24(24):2920–2925. http://dx.doi.org/10.1016/j.cub.2014.10.038

Nadrowski B, Albert JT, Gopfert MC (2008) Transducer-based force generation explains active process in Drosophila hearing. Curr Biol 18(18):1365–1372. https://doi.org/10.1016/j.cub.2008.07.095

Thurm U, Erler G, Godde J, Kastrup H, Keil T, Volker W et al (1983) Cilia specialized for mechanoreception. J Submicrosc Cytol Path 15(1):151–155

Tsubouchi A, Caldwell Jason C, Tracey WD (2012) Dendritic filopodia, ripped pocket, NOMPC, and NMDARs contribute to the sense of touch in Drosophila larvae. Curr Biol 22(22):2124–2134. http://dx.doi.org/10.1016/j.cub.2012.09.019

Walker RG, Willingham AT, Zuker CS (2000) A Drosophila mechanosensory transduction channel. Science 287(5461):2229–2234. https://doi.org/10.1126/science.287.5461.2229

Yan Z, Zhang W, He Y, Gorczyca D, Xiang Y, Cheng LE et al (2013) Drosophila NOMPC is a mechanotransduction channel subunit for gentle-touch sensation. Nature 493(7431):221–225. https://doi.org/10.1038/nature11685

Yorozu S, Wong A, Fischer BJ, Dankert H, Kernan MJ, Kamikouchi A et al (2009) Distinct sensory representations of wind and near-field sound in the Drosophila brain. Nature 458(7235):201–205. http://www.nature.com/nature/journal/v458/n7235/suppinfo/nature07843_S1.html

Zhang W, Yan Z, Li B, Jan LY, Jan YN (2014) Identification of motor neurons and a mechanosensitive sensory neuron in the defecation circuitry of Drosophila larvae. elife 3. https://doi.org/10.7554/eLife.03293

Chapter 5
Drosophila Mechanotransduction Channels

Abstract In Chap. 4, we discussed the working mechanisms of several fly mecha-noreceptors. These mechanoreceptors have developed striking structural-mechani-cal features that couple the environmental signals to the proximity of the mechanosensory cells. These proximal stimuli are then transformed by a transduc-tion apparatus, often associated with the cell membrane, to the electrochemical sig-nals in cells, which then initiates the neuronal impulses. In this chapter, we will discuss the core components of mechanotransduction apparatus at the molecular level, including the mechanotransduction channel and the gating spring. Two bona fide mechanotransduction channels are identified in *Drosophila melanogaster*, i.e., NompC/TRPN and DmPiezo. One molecular candidate for gating spring is identi-fied, i.e., the ankyrin-repeat domain of NompC. In this chapter, we will introduce the structure, function, and physiology of these molecules and discuss their working mechanisms in fly mechanoreceptors.

5.1 Overview

In Chap. 2, we introduced the "gating-spring" model. Here, we summarize the theo-retical background of the transduction apparatus in another way. Consider that if the *e*-fold change in the opening probability of the transduction channel is associated with a length change, Δl, of the gating spring (stiffness: κ) in the direction of force (*f*), then the open/closed probability ratio will change by a factor of

$$\exp^{\frac{-f\Delta l}{kT}}$$

(*k* is the Boltzmann constant and *T* is absolute temperature)

   Therefore, we obtain $kT = f\Delta l = 0.5\kappa\Delta l^2 \approx 4$ pN$\square$nm (Howard 2001). This pre-dicts that a small κ will be accompanied by a relatively large Δl. If κ is 0.5 pN/nm, then Δl is 4 nm. In this case, every 4 nm change in the length of the gating spring leads to the *e*-fold change in the opening probability of the transduction channel. If κ is 8 pN/nm, then Δl is 1 nm, namely, that the 1 nm deformation of the spring is able to cause *e*-fold change in the channel's opening probability. This suggests that

© The Author(s) 2017
X. Liang et al., *Mechanosensory Transduction in Drosophila Melanogaster*,
SpringerBriefs in Biochemistry and Molecular Biology,
DOI 10.1007/978-981-10-6526-2_5

when the gating spring is stiffer, the channel is more sensitive to the deformation. The dynamic range of the channel depends on κ as well. Let's assume a closed state ($P_0 = 0.001$) and an open state ($P_0 = 1.0$) of the transduction channel, 10^3 (e^7) folds change in P_0 of the two functional states. When κ is 0.5 pN/nm, the dynamic range of stimulating deformation is 28 nm and it reduces to 7 nm if κ is 8 pN/nm. This shows that if the transduction channel is coupled to a stiffer spring, then it has a smaller dynamic range. With these calculations, we confirm the conclusions in Chap. 2 that the stiffness of the mechaincal signaling pathway (i.e., the gating spring) shapes the performances of a transduction apparatus, in terms of the sensitivity and the sensory dynamic range.

In this view, it is likely that each transduction channel, based on its surrounding organization in cells and its physiological stimuli (touch or sound), is coupled to a specific molecular spring. The structural and mechanical studies on the mechanotransduction channel can lead us to the understanding of the structures and functions of the mechanotransduction channels.

5.2   No Mechanoreceptor Potential C (NompC)

5.2.1   Overview on Fly NompC

Roughly 20 years ago, Kernan et al. in Zuker lab started to use fly larvae as the model organism to screen for the mechanoreceptive mutants because the adult fly mutants for the external mechanosensory organs are difficult to recover due to the low viability and severe uncoordination phenotype. A set of mechanoreception-related genes were identified in this screen, including *nompC* (Kernan et al. 1994). *nompC* mutant larvae show a defective response to touches, and the adults show a complete loss of the mechanoreceptor potential in the bristle mechanoreceptors. Later, Walker et al., also from the Zuker lab, reported that *nompC* mutants abolish mechanosensory signaling (Walker et al. 2000). In this study, three prematured *nompC* nonsense mutants (*nompC¹*, *nompC²*, *nompC³*) showed the nearly complete loss of mechanosensory current and one missense point mutation strain (*nompC⁴*) showed a faster adaptation kinetics in the electrophysiological recording experiments on the bristle mechanoreceptors. Based on the sequence analysis and the expression pattern of NompC, Walker et al. proposed that NompC is a candidate mechanotransduction channel in fly mechanoreceptors. A few year later, the correct isoform of NompC was identified and cloned in a study by Cheng et al. in the Jan lab (UCSF) in which they reported that NompC is required in the locomotion of both larvae and adult flies (Cheng et al. 2010). In 2013, Yan et al. in the Jan lab reported that NompC is a bona fide mechanotransduction channel and is required in the gentle touch sensation of fly larvae (Yan et al. 2013). Most recently (2017), a striking work from the collaboration of the Jan lab and Yifan Cheng's lab (UCSF) reported the atomic structure of NompC using revolutionized cyroEM technique

(Jin et al. 2017), which opens now the gate to study the structural mechanics and gating mechanism of NompC in greater details. Except for the basic characterization of NompC listed above, there are many important findings on the physiological functions of NompC in the mechanoreceptors of *Drsosphila* and other organisms (e.g., *C. elegans* and *zebra fish*) (Zhang et al. 2013, 2015; Liang et al. 2013; Lehnert et al. 2013; Effertz et al. 2012; Liang et al. 2011; Effertz et al. 2011; Lee et al. 2010; Kang et al. 2010; Gopfert et al. 2006). These findings cover a broad range of topics and are in great depth, which we will review in the following section.

5.2.2   Structure of NompC

The early sequence analysis showed that NompC has a low but significant similarity with the TRP (transient receptor potential) ion channel superfamily (Walker et al. 2000). It is the only member in the TRPN family. The recently resolved atomic structure of NompC reveals the overall architecture of NompC in greater details (Jin et al. 2017). In this study, Jin et al. showed that NompC forms a homotetrameric ion channel. The structure of NompC is highly modulized into three major parts, i.e., the transmembrane (TM) domains, the neck region, and the ankyrin-repeat (AR) domain, together with the largely unstructured C- and N-terminal sequences (Fig. 5.1a).

First, each NompC subunit contains a set of six transmembrane (TM) α-helices (S1–S6). Four TM modules form a homotetramer through the domain swap interactions (Fig. 5.1b). An S4–S5 linker is found on the cytoplasmic side and is nearly in parallel with the lipid bilayer (the green helix in Fig. 5.1b). The pore loop is located between S5 and S6 and appears to be a reentrant structure (the blue helix in Fig. 5.1b). These structures are similar to the other TRP channels and the voltage-gated potassium channels. NompC has a novel and unique TM part called the "pre-S1 elbow" (the red helices in Fig. 5.1b). As shown in its name, it locates in the front of S1. This structure contains two helices and spans only half of the lipid bilayer from the cytoplasmic side to form an inverted "V" shape. This domain has an extensive interaction with the S1 helix. The N-terminal of "pre-S1 elbow" connects to the linker helices in the neck region.

Second, the neck region (Fig. 5.1d) is formed of a network of interactions between a stack of linker helices, the TRP domain and the C-terminus (red in Fig. 5.1d). The linker helices are a set of helices stacked together between the AR domain and the TM domains (light blue in Fig. 5.1d). They form a linking structure that connects the S4–S5 linker via the TRP domain (green in Fig. 5.1d) to the AR domain. The way that the linker helices organize is similar to the stacked helices in TRPA1 but distinct from the known structures of the TRPV channels.

Third, the most striking feature of NompC is a quadruple ankyrin-repeat (AR) helices (referred to as "AR bundle" from here on) (Fig. 5.1c). Each AR helix resembles a helical spring, which agrees well with the previously predicted structure. The AR helices in the bundle interact with the adjacent subunits through four contact

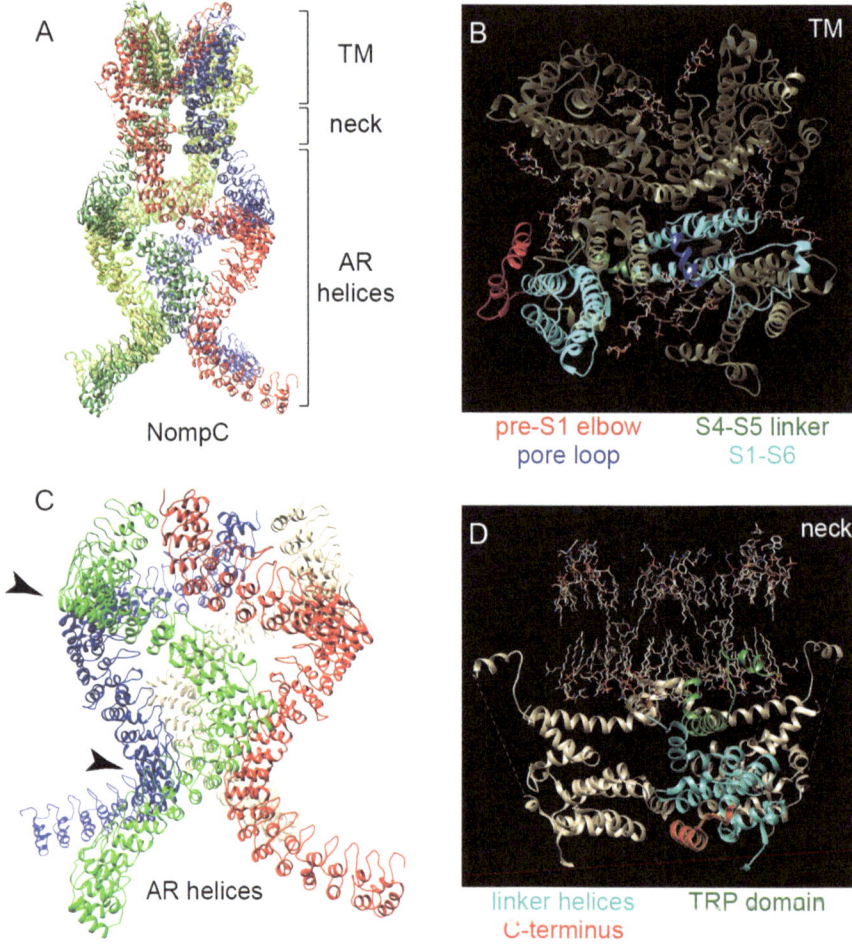

Fig. 5.1 The cryoEM structure of NompC (based on a PDB model, accession number 5VKQ) (Jin et al. 2017). (**a**) The overall architecture of homotetrameric NompC includes three parts (*side view*): the transmembrane domains, the neck region, and the quadruple AR helices. (**b**) Detailed structure of the transmembrane domains (*top view*). Different structural elements are labeled in different colors as indicated. Note that the transmembrane domains of only one in the four subunit are highlighted and the others are labeled in *light yellow*. (**c**) Four ankyrin helices form a bundle through two types of contacts as indicated by the *black arrowheads* (*side view*). (**d**) Detailed structure of the neck region (*side view*). Different structural motifs are labeled in different colors as indicated (Note that only one subunit is highlighted and the other three are in *light yellow*. The lipid densities are included to show the position of the transmembrane domains)

sites (indicated by the arrowheads in Fig. 5.1c). These four contacts sites are AR24-26, AR16-18, AR7-10, and AR9-12, in which the AR24-26 of one subunit interacts with the AR16-18 of its left neighbor while the AR7-10 interacts with the AR9-12. Therefore, the four contact sites form two types of constrictions along the length of the AR helices.

Fig. 5.2 Two types of forces that might gate the NompC channel (*Ex* extracellular, *In* intracellular, *f* the normal force from the AR helices to the channel gate through the neck region, *T* the lateral tension from the lipid bilayer surrounding the transmembrane domains of NompC)

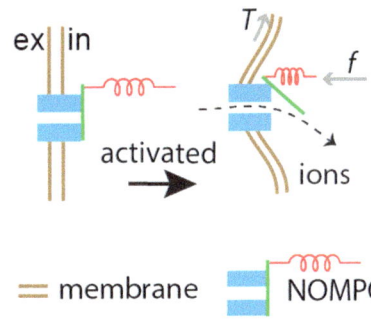

The structure of the N-terminus of NompC is predicted to be unstructured and remains to be determined. The C-terminus of NompC is largely unstructured but contains several helical structures that interact with the linker helices as a part of the neck region.

5.2.3   Gating of NompC

The gating mechanism of NompC is yet to be determined in details. We discuss here two possible force-gating models of NompC.

Two models are proposed for the gating mechanisms of a mechanotransduction channel (Fig. 5.2). The first one is the "lateral membrane tension" model in which the proximal stimulus for the force-sensitive channel is the tension from the lateral lipid bilayer (Zanini and Gopfert 2013). A typical example for this model is the Msc channels in bacterial cells (Kung et al. 2010). The second model is the "tethered model" (Zanini and Gopfert 2013; Gillespie and Walker 2001) in which the transduction channels are tethered to either the intracellular cytoskeleton or the extracellular matrix via macromolecular linkers. In the tethered model, transduction channels receive forces through the linkers. Because the normal forces through the linker cause the membrane deformation which also creates the lateral membrane tension, it is difficult to definitively completely rule out one or the other. Therefore, membrane tension or force through the tether-it is still a question.

Another way to think about the differences in these two models is to consider whether the pore-forming domain is switched on/off though the protein-lipid interaction or intramolecular interaction among different domains in the protein structure. Here, we take advantage of the NompC structure to discuss these two possible models.

At the first glance, the overall shape of the tetrameric NompC gets along well with the "tethered model" (Jin et al. 2017). One unique feature of NompC is that its AR bundle mediates the interaction between NompC and the microtubule in cells (Jin et al. 2017; Liang et al. 2013; Cheng et al. 2010). When NompC is pushed/pulled toward or away from the microtubules, the compressive/stretching forces through the AR helices will arrive at the channel domain of NompC (Fig. 5.2). On one hand, the AR helices are coupled with the TM domains through the linker helices, C-terminus, and the TRP domain. The TRP domain is a key linker. On one side, it is associated with the AR helices through the pre-S1 elbow and the linker helices. On the other side, the TRP domain extends directly to the C-terminus of S6 and is packed closely to the S4–S5 linker. These TM domains are potentially gating-related regions. The direct intramolecular linkage provides a structural basis for the force transmitted from the AR bundle to the gate and to bias the opening probability of the transduction channel. In fact, such a direct association of the TRP domain to the channel gate is also observed in other TRP channels, e.g., TRPV1 (Liao et al. 2013), which suggests a potentially conservative role of the TRP domain in the TRP channel family. By comparing three substructures of NompC from the cryoEM dataset, it was found that while the AR bundle is slightly moved away from the TM domains, the linker domains also ungo an anticlockwise rotation, suggesting that the linker domains might act as the final adaptor for the fine tuning (e.g., by adjusting the force direction or intensity) of the gating signal for the pore-forming domain of the channel. These structural details support the "tethered model" and the proximal stimulus being the normal forces from the linkers.

However, other structural details suggest that the potential roles of lateral membrane tension cannot be completely excluded. First, there is tightly bound lipid density associated with the hydrophobic cleft formed by the TM domains, including the regions that are close to the channel gate (e.g., pre-S1 elbow, S1, S4, S4–S5 linker). Second, His1423 in the S4–S5 linker region is involved in the lipid-protein interaction, and the His1423Ala mutant channel shows no mechanosensory responses (Jin et al. 2017). Third, the density of lipid is present in the cryoEM data sets collected using different sample preparation methods, suggesting a robust lipid-protein interaction that is likely present in the native conditions. Taken together, these observations show a functionally important lipid-protein interaction in channel gating and thereby suggest the potential roles of lateral membrane tension in channel gating.

In summary, the normal forces through the AR helices and lateral tension from lipid membrane are both present in the proximal surroundings of NompC in cells. To differentiate the proximal trigger for the channel gating needs to capture the dynamic conformational changes associated with the channel gating. While this may be technically challenging for the experimentalists, the in silico simulations using the the molecular dynamics method might help on this issue.

5.2.4 The Gating Spring of NompC

Despite being proposed for more than 30 years, gating spring remains to be a conceptual element in the mechanotransduction apparatus. Its molecular identity has not yet been confirmed in any of the model systems. The "AR bundle" of NompC is the most intriguing molecular candidate for gating spring based on two lines of evidences (Fig. 5.1c).

The first set of evidence is from the AR bundle mechanics. When NompC was first identified in Kernan's screening and in Walker's functional studies, an exceptional feature found in this molecules is the N-terminal 1150 amino acid residues that consist of 29 ankyrin repeats (AR). AR is a 33-residue protein motif that mediates specific protein-protein interactions with a wide range of targets, particularly involved in connecting the plasma membrane and the intracellular cytoskeletal network. The AR domain of NompC attacts an extensive attension because it is considered as a candidate for the molecular spring. First, based on the mechanical measurements and the helical structure of the 12-ankyrin-repeat domain from ankyrin, it was predicted that the AR domain of NOMPC (29 ARs) forms one turn of a helix (Howard and Bechstedt 2004). The stiffness, estimated from the geometry of the helix and assuming that the constituent polypeptides have a similar Young's modulus to actin or tubulin (2 GPa), is on the order of 1 pN/nm (Howard and Bechstedt 2004). Second, based on the molecular dynamics simulation on the AR domains of ankyrin-R and TRPA1, it was calculated that the AR domain of NompC has a stiffness on the order of 4 pN/nm (Sotomayor et al. 2005), similar to what was estimated (Howard and Bechstedt 2004). Third, Lee et al. with AFM studies of a 24-AR domain showed a stiffness of the full-length domain or roughly 2 pN/nm (Lee et al. 2006), further confirming that the 29 ARs of NompC form a compliant structure. These mechanical experiments and estimations provide the first set of evidence that support the hypothesis that the AR domain of NompC acts as the gating spring.

The cryoEM structure of NompC shows a striking quadruple AR helix architecture at the N-terminal half of the homotetramer NompC (Jin et al. 2017). As described above, the four AR subunits interact with their neighbors through two types of contacts (Fig. 5.3a). This new model allows us to to consider the compound

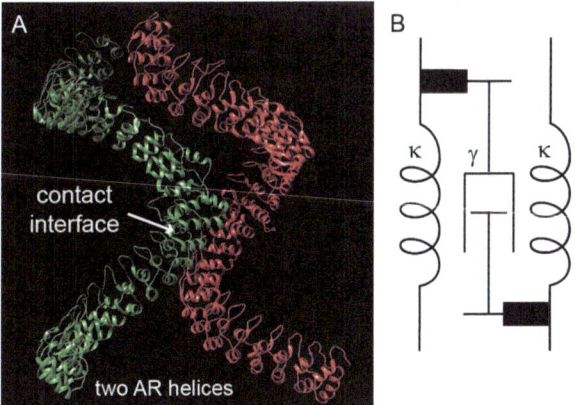

Fig. 5.3 The direct contact between the adjacent AR helices (**a**) might be considered as the frictional interaction generated by the relative slip motion of one helix in relative to the other (**b**). The contact interface is indicated by a *white arrow* in panel (**a**). In panel (**b**), two springs represent two AR helices, and the dashpot is used to represent the frictional interaction between two helices

stiffness of the tetrameric AR domains from a structural point of view. If the inter-subunit interaction is very strong, the four AR helices are "glued" together, and the AR helices would have to always move synchronously. In this case, given the stiffness of a single AR helix is around 1 pN/nm, the tetrameric AR helices would have a stiffness of about 4 pN/nm, still in the range of being a gating spring for the transduction channel. The new structure shows that the interfaces of both types of interactions are relatively broad (spanning several ARs) and are formed by an array of polar residues in which the several possible interacting partners on adjacent subunits are present. This suggests that the inter-subunit contacts between the adjacent subunits are not fixed but flexible to some extent. How to consider the effects of the inter-subunit interactions of the mechanics of the AR helices? One possibility is that the inter-subunit links are frictional (Fig. 5.3b), namely, that the AR domains can slide with respect to one another within the AR helices. In this case, the movement of the AR helices can be considered as the motion of a spring with damping. If the inter-subunit friction is relatively large, the length change of the AR helices under the stimulating stretch or compression is expected to be slowed down but still be able to reach the full range (Howard 2001). Therefore, the new quadruple AR helices architecture also supports the AR helices to be the gating spring.

The second set of evidence is from the observation that the AR domain of NompC contributes the membrane-microtubule connector (MMC) in the mechanoreceptive organelle of fly campaniform mechanoreceptor (see Chap. 4, Sect. 4.3.3) (Liang et al. 2013; Voelker 1982). In addition, early analysis of the geometry of the campaniform receptor suggested that compression or stretch of the MMC opens the channel (Liang et al. 2013; Voelker 1982). Transmission EM micrographs of mechanically stimulated receptors indicate that the MMCs might be stretched or compressed as much as several nanometers. Furthermore, mechanical measurements in fly bristle receptors, which are structurally related to campaniform receptors, indicate that the stiffness of an individual MMC is on the order of 3 pN/nm (Thurm et al. 1983), similar to what is estimated for the AR bundle. These in vivo observations show that the AR bundle in NompC homotetramer localizes at the right place to convey the force to the transduction channel and thereby also support the hypothesis that the AR bundle is the gating springs for NompC.

There are still a number of experiments required to definitely demonstrate that the AR bundle acts as the gating spring, including the experimental measurement of the AR bundle stiffness at the single-molecule level and the contribution of the AR bundle to the mechanical rigidity of the mechanoreceptive organelle in vivo. These measurements are challenging to the experimentalists in the next few years. However, the efforts made in the structural and mechanical analysis of the NompC AR bundle reveal the first molecular picture for the conceptual gating-spring theory and a paradigm for the tethered model.

5.2.5   NompC-Microtubule Interaction

One unique feature of NompC is that it directly binds to microtubules. This is well evidenced by a number of observations. First, the overexpressed NompC in the heterologous system co-localizes with the microtubules in cells (Cheng et al. 2010). Second, the purified NompC molecules bind to in vitro assembled microtubules (Jin et al. 2017; Zhang et al. 2015). Third, the AR helices contribute to the membrane-microtubule connectors in vivo, and the size of the AR helices is large enough to span the gap between the membrane and the microtubule (Liang et al. 2013). The direct NompC-microtubule interaction provides a linkage for the force transmission from the rigid microtubules via a compliant linker to the transduction channels (Liang et al. 2014). However, it is not yet clear how NompC binds to the microtubules. An interesting point is that the four feet of the AR helices have a fourfold rotational symmetry (Fig. 5.4a), while the microtubule lattice has a translational symmetry (Fig. 5.4b). The question is how to match these two different symmetries when NompC directly binds to the microtubule. The N-terminus of NompC may play an important role as a structural adaptor. However, the structure of the N-terminal sequence has not yet been resolved in the cryoEM structure and is predicted to be unstructured. One possibility is that the N-termini of NompC may become structured once they interact with the microtubule. More structural information is clearly required here.

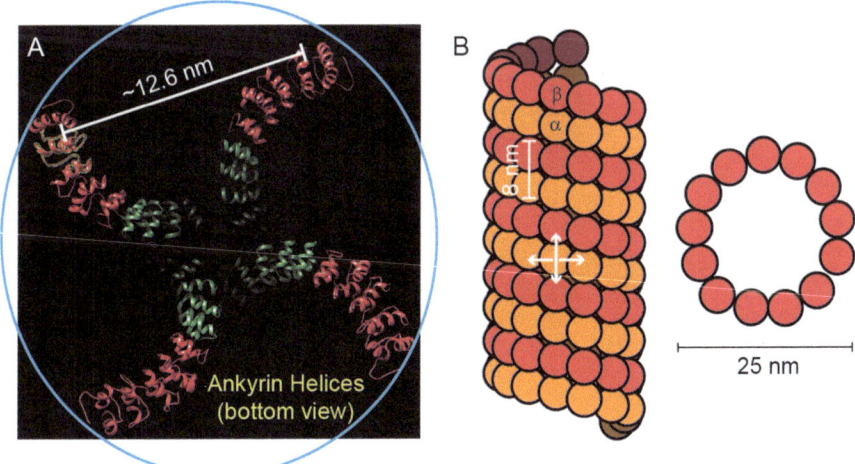

Fig. 5.4 The quadruple AR helices have fourfold rotational symmetry (**a**), and the microtubule lattice has a translational symmetry (**b**) with the longitudinal unit length of 8 nm (*left panel* in **b**) and the circumferential unit length of about 6 nm (*right panel* in **b**). It is a question how to match the rotational symmetry of the AR helices bundle to the translational symmetry of the microtubule wall

5.2.6 *Physiological Roles of Fly NompC*

NompC is found expressed in various mechanosensation-related cells and involved in different mechanosensory processes. We summarize the physiological roles of NompC in Table 5.1. The gating mechanisms of NompC in these cells are mostly unclear. It is expected that NompC might always work together with microtubules, just like in the campaniform and bristle mechanoreceptors. However, such NompC-microtubule complex has not been observed yet in other cells, e.g., the Johnston's organ sensory cells and the class III da neurons. Therefore, more mechanistic studies on these mechanosensory cells are required to further elucidate the gating mechanisms of NompC in different sensory cells.

Table 5.1 The physiological roles of NompC/TRPN

Species	Tissues or cell types		Localization	Physiological functions
Drosophila melanogaster	Type I	Bristle sensilla	Bristles on the fly body and proboscis	Touch
		Campaniform sensilla	Wing bases and halteres	Cuticle deformation
		Scolopidial organs	Johnston's organs in the antennae, leg joints and chordotonal organs in the larvae	Hearing, locomotion
	Type II	Class I da neurons	Larval body wall	Locomotion
		Class III da neurons	Larval body wall	Gentle touch
		Proprioceptive neurons	Leg joint	Proprioception
Caenorhabditis elegans	CEP neuron, DVA neuron		Nose tip of the worm	Stretch
Danio rerio	Hair cells		Inner ear in the embryo and larva	Hearing
Xenopus laevis	Hair cells, some epithelial cells		Lateral line and vestibular system, embryonic and larval embryonic epidermis	Possibly hearing

References for Table 5.1
1. Walker, R.G., A.T. Willingham, and C.S. Zuker, A Drosophila mechanosensory transduction channel. Science, 2000. 287(5461): p. 2229–34.
2. Kim, J., et al., A TRPV family ion channel required for hearing in Drosophila. Nature, 2003. 424(6944): p. 81–4.

(continued)

Table 5.1 (continued)

3. Sidi, S., R.W. Friedrich, and T. Nicolson, NompC TRP channel required for vertebrate sensory hair cell mechanotransduction. Science, 2003. 301(5629): p. 96–9.

4. Shin, J.B., et al., Xenopus TRPN1 (NOMPC) localizes to microtubule-based cilia in epithelial cells, including inner-ear hair cells. Proc Natl Acad Sci U S A, 2005. 102(35): p. 12572–7.

5. Gopfert, M.C., et al., Specification of auditory sensitivity by Drosophila TRP channels. Nat Neurosci, 2006. 9(8): p. 999–1000.

6. Li, W., et al., A C. elegans stretch receptor neuron revealed by a mechanosensitive TRP channel homologue. Nature, 2006. 440(7084): p. 684–7.

7. Kang, L., et al., C. elegans TRP family protein TRP-4 is a pore-forming subunit of a native mechanotransduction channel. Neuron, 2010. 67(3): p. 381–91.

8. Lee, J., et al., Drosophila TRPN(=NOMPC) channel localizes to the distal end of mechanosensory cilia. PLoS One, 2010. 5(6): p. e11012.

9. Effertz, T., R. Wiek, and Martin C. Göpfert, NompC TRP Channel Is Essential for Drosophila Sound Receptor Function. Current Biology, 2011. 21(7): p. 592–597.

10. Liang, X., et al., NOMPC, a member of the TRP channel family, localizes to the tubular body and distal cilium of Drosophila campaniform and chordotonal receptor cells. Cytoskeleton (Hoboken), 2011. 68(1): p. 1–7.

11. Effertz, T., et al., Direct gating and mechanical integrity of Drosophila auditory transducers require TRPN1. Nat Neurosci, 2012. 15(9): p. 1198–1200.

12. Tsubouchi, A., Jason C. Caldwell, and W.D. Tracey, Dendritic Filopodia, Ripped Pocket, NOMPC, and NMDARs Contribute to the Sense of Touch in Drosophila Larvae. Current Biology, 2012. 22(22): p. 2124–2134.

13. Lehnert, B.P., et al., Distinct roles of TRP channels in auditory transduction and amplification in Drosophila. Neuron, 2013. 77(1): p. 115–28.

14. Liang, X., et al., A NOMPC-dependent membrane-microtubule connector is a candidate for the gating spring in fly mechanoreceptors. Curr Biol, 2013. 23(9): p. 755–63.

15. Yan, Z., et al., Drosophila NOMPC is a mechanotransduction channel subunit for gentle-touch sensation. Nature, 2013. 493(7431): p. 221–5.

16. Zhang, W., et al., Sound response mediated by the TRP channels NOMPC, NANCHUNG, and INACTIVE in chordotonal organs of Drosophila larvae. Proc Natl Acad Sci U S A, 2013. 110(33): p. 13612–7.

17. Cheng, L.E., et al., The role of the TRP channel NompC in Drosophila larval and adult locomotion. Neuron, 2010. 67(3): p. 373–80.

18. Chadha, A., M. Kaneko, and B. Cook, NOMPC-dependent mechanotransduction shapes the dendrite of proprioceptive neurons. Neurosci Lett, 2015. 597: p. 111–6.

19. Ramdya, P., et al., Mechanosensory interactions drive collective behaviour in Drosophila. Nature, 2015. 519(7542): p. 233–6.

20. Sanchez-Alcaniz, J.A., et al., A mechanosensory receptor required for food texture detection in Drosophila. Nat Commun, 2017. 8: p. 14192.

5.2.7   NompC/TRPN in Other Organisms

Although NompC is mostly studied in *Drosophila melanogaster*, it is also found in both invertebrates (e.g., fly and worm) and lower vertebrates (e.g., fish) but not in the higher vertebrates and mammals (Fig. 5.5 and Appendix). The *C. elegans* homologue of NompC, TRP-4, expresses in the ciliated mechanosensory neurons, similar to the fly NompC. TRP-4 is required for the mechanosensory conductance in these ciliated sensory neurons, and the point mutations in the predicted pore-forming

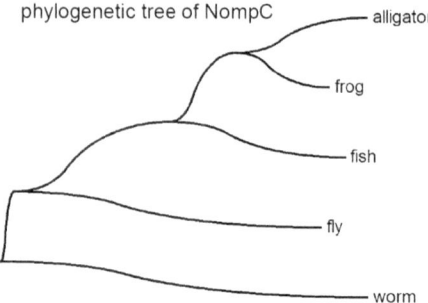

Fig. 5.5 The phylogenetic tree of NompC/TRPN (using MEGA7). NompC is present in invertebrate and lower vertebrate but missing in higher vertebrates and mammals. The sequence alignment of these five NompC homologues is attached at the end of this chapter (Appendix, Kumar et al. 2016; Waterhouse et al. 2009)

domain of TRP-4 alter the channel properties (Kang et al. 2010). NompC homologues are also found in *zebra fish* and in *Xenopus* (Shin et al. 2005; Sidi et al. 2003), but the working mechanisms of NompC in these organisms still remain elusive. The sequence similarity between NompC homologues in various organisms is relatively high (Appendix), suggesting these homologues might function in a similar ways as fly NompC (i.e., in a NompC-microtubule complex and as an ion channel). One interesting point is that the N-termini of these NompC homologues, thought to contain the microtubule binding domain of NompC, are highly divergent with only a few short conserved regions. It would be interesting to investigate how these divergent regions can all mediate the microtubule binding of different NompC homologues.

5.3   DmPiezo

The second bona fide mechanosensitive channel found in *Drosophila melanogaster* is DmPiezo (Volkers et al. 2015). Piezo proteins, Piezo 1 and Piezo 2, were first identified in a mammalian neuroblastoma cell line using the siRNA-based screening strategy (Volkers et al. 2015; Coste et al. 2010, 2012). Later, Piezo proteins were found to be mechanosensitive pore-forming ion channel that displays cationic nonselective and mechanosensitive currents (Coste et al. 2012). Since their identification, Piezo molecules have been found to be widely involved in many physiological processes in both sensory and non-sensory cells (Volkers et al. 2015). A medium-resolution structure of mouse Piezo1 has been resolved using the cryoEM technique (Ge et al. 2015). This structure shows that the Piezo proteins are homotrimer, which has two striking features: (1) a large transmembrane domain with 18

Fig. 5.6 The phylogenetic tree of Piezo (using MEGA7). Piezo proteins are present in various organisms of different evolutional levels, suggesting its conserved functions across the evolution (Kumar et al. 2016; Waterhouse et al. 2009)

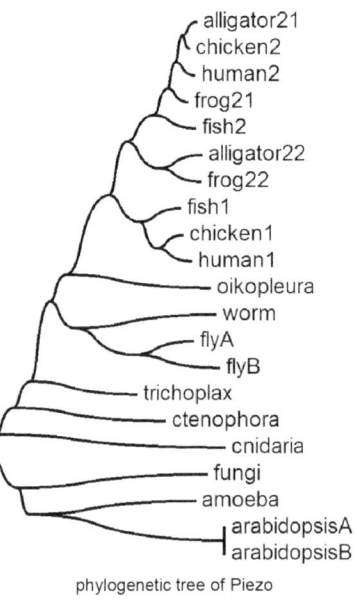

phylogenetic tree of Piezo

transmembrane segments for each subunit, and (2) a large extracellular domain with three propeller-shaped structure. The revelation of this structure is important for the ongoing and future revelation of the gating mechanism of Piezo as well as its regulations by the associated molecules and drugs. Piezo proteins are found in many different organisms (Fig. 5.6), ranging from the lower animals (e.g., oikopleura and amoeba) to mammals (e.g., human), suggesting its critical and fundamental roles in general physiology. Although most of the characterization works on Piezo have been done on mammalian Piezo proteins and have been reviewed elsewhere (Volkers et al. 2015), there are a number of recent findings on the physiological functions of the DmPiezo: (1) DmPiezo was found expressed in sensory neurons and some non-neuronal tissues (Coste et al. 2010); (2) as described in Chap. 4, DmPiezo contributes to one of the two signaling pathways in mediating the nociception in class IV da sensory neurons (Mauthner et al. 2014; Guo et al. 2014; Gorczyca et al. 2014; Kim et al. 2012); and (3) DmPiezo is involved in the stretch-activated mechanotransduction in the dorsal bipolar dendritic sensory neurons (Suslak et al. 2015). Future challenges to the experimentalists on Piezo and DmPiezo are to understand their gating mechanisms in greater details, to improve the resolution in the structure to see more structural details, and to further explore their physiological functions.

Appendix

Sequence alignment of five NompC homologues. The alignment of the sequences and the construction of the phylogenetic tree are carried out by the MEGA (version 7) software suite. The sequences are aligned by the MUSCLE algorithm with default parameters. The aligned sequences are visualized using Jalview software.

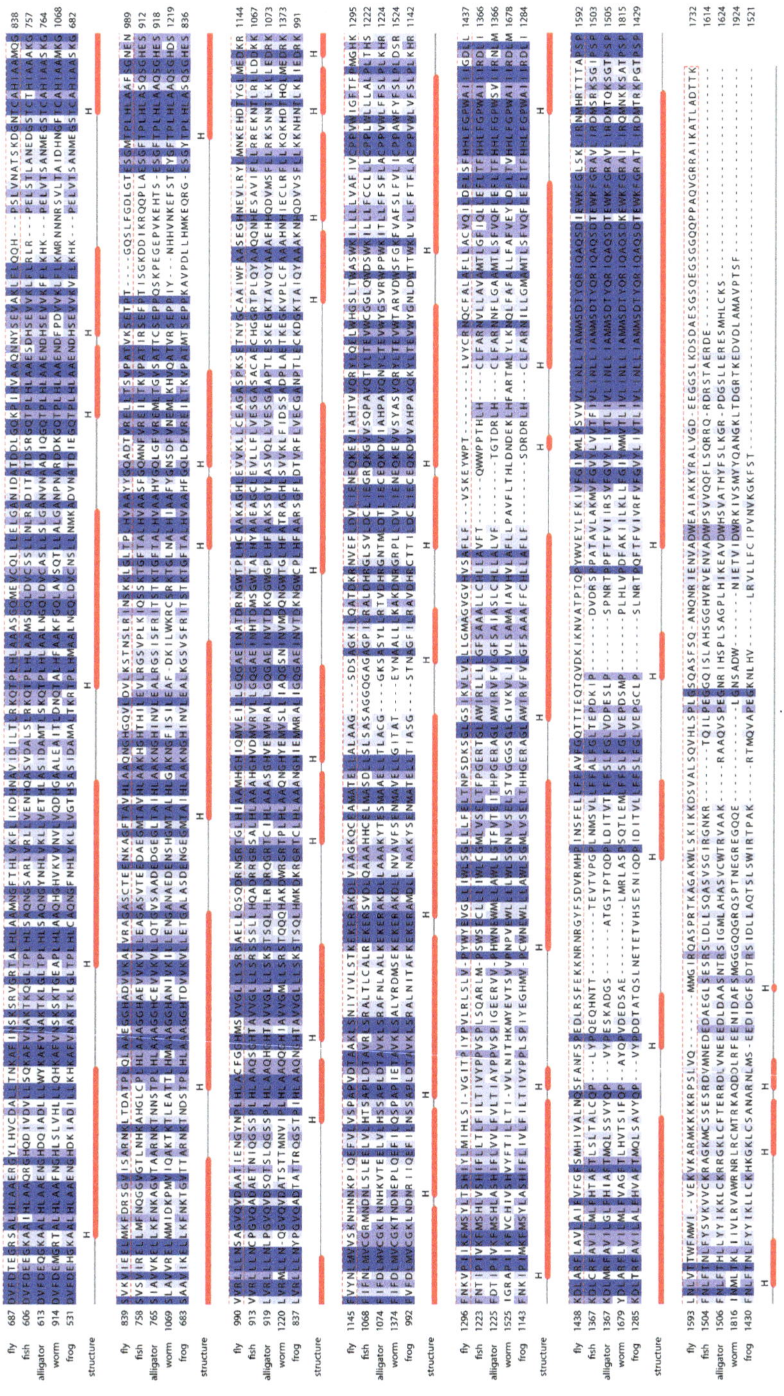

References

Cheng LE, Song W, Looger LL, Jan LY, Jan YN (2010) The role of the TRP channel NompC in Drosophila larval and adult locomotion. Neuron 67(3):373–380. https://doi.org/10.1016/j.neuron.2010.07.004

Coste B, Mathur J, Schmidt M, Earley TJ, Ranade S, Petrus MJ et al (2010) Piezo1 and Piezo2 are essential components of distinct mechanically activated cation channels. Science 330(6000):55–60. https://doi.org/10.1126/science.1193270

Coste B, Xiao B, Santos JS, Syeda R, Grandl J, Spencer KS et al (2012) Piezo proteins are pore-forming subunits of mechanically activated channels. Nature 483(7388):176–181. https://doi.org/10.1038/nature10812

Effertz T, Wiek R, Göpfert MC (2011) NompC TRP channel is essential for Drosophila sound receptor function. Curr Biol 21(7):592–597. http://dx.doi.org/10.1016/j.cub.2011.02.048

Effertz T, Nadrowski B, Piepenbrock D, Albert JT, Gopfert MC (2012) Direct gating and mechanical integrity of Drosophila auditory transducers require TRPN1. Nat Neurosci 15(9):1198–1200. http://www.nature.com/neuro/journal/v15/n9/abs/nn.3175.html#supplementary-information

Ge J, Li W, Zhao Q, Li N, Chen M, Zhi P et al (2015) Architecture of the mammalian mechanosensitive Piezo1 channel. Nature 527(7576):64–69. https://doi.org/10.1038/nature15247

Gillespie PG, Walker RG (2001) Molecular basis of mechanosensory transduction. Nature 413(6852):194–202. https://doi.org/10.1038/35093011

Gopfert MC, Albert JT, Nadrowski B, Kamikouchi A (2006) Specification of auditory sensitivity by Drosophila TRP channels. Nat Neurosci 9(8):999–1000. https://doi.org/10.1038/nn1735

Gorczyca David A, Younger S, Meltzer S, Kim Sung E, Cheng L, Song W et al (2014) Identification of Ppk26, a DEG/ENaC channel functioning with Ppk1 in a mutually dependent manner to guide locomotion behavior in Drosophila. Cell Rep 9(4):1446–1458. http://dx.doi.org/10.1016/j.celrep.2014.10.034

Guo Y, Wang Y, Wang Q, Wang Z (2014) The role of PPK26 in Drosophila larval mechanical nociception. Cell Rep 9(4):1183–1190. http://dx.doi.org/10.1016/j.celrep.2014.10.020

Howard J (2001) Mechanics of motor proteins and the cytoskeleton. Sinauer Associates, Publishers, Sunderland

Howard J, Bechstedt S (2004) Hypothesis: a helix of ankyrin repeats of the NOMPC-TRP ion channel is the gating spring of mechanoreceptors. Curr Biol 14(6):R224–R226. https://doi.org/10.1016/j.cub.2004.02.050

Jin P, Bulkley D, Guo Y, Zhang W, Guo Z, Huynh W et al (2017) Electron cryo microscopy structure of the mechanotransduction channel NOMPC. Nature. https://doi.org/10.1038/nature22981

Kang L, Gao J, Schafer WR, Xie Z, Xu XZ (2010) C. Elegans TRP family protein TRP-4 is a pore-forming subunit of a native mechanotransduction channel. Neuron 67(3):381–391. https://doi.org/10.1016/j.neuron.2010.06.032

Kernan M, Cowan D, Zuker C (1994) Genetic dissection of mechanosensory transduction: mechanoreception-defective mutations of Drosophila. Neuron 12(6):1195–1206

Kim SE, Coste B, Chadha A, Cook B, Patapoutian A (2012) The role of Drosophila Piezo in mechanical nociception. Nature 483(7388):209–212. https://doi.org/10.1038/nature10801

Kumar S, Stecher G, Tamura K (2016) MEGA7: molecular evolutionary genetics analysis version 7.0 for bigger datasets. Mol Biol Evol 33(7):1870–1874. https://doi.org/10.1093/molbev/msw054

Kung C, Martinac B, Sukharev S (2010) Mechanosensitive channels in microbes. Annu Rev Microbiol 64:313–329. https://doi.org/10.1146/annurev.micro.112408.134106

Lee G, Abdi K, Jiang Y, Michaely P, Bennett V, Marszalek PE (2006) Nanospring behaviour of ankyrin repeats. Nature 440(7081):246–249. https://doi.org/10.1038/nature04437

Lee J, Moon S, Cha Y, Chung YD (2010) Drosophila TRPN(=NOMPC) channel localizes to the distal end of mechanosensory cilia. PLoS One 5(6):e11012. https://doi.org/10.1371/journal.pone.0011012

Lehnert BP, Baker AE, Gaudry Q, Chiang AS, Wilson RI (2013) Distinct roles of TRP channels in auditory transduction and amplification in Drosophila. Neuron 77(1):115–128. https://doi.org/10.1016/j.neuron.2012.11.030

Liang X, Madrid J, Saleh HS, Howard J (2011) NOMPC, a member of the TRP channel family, localizes to the tubular body and distal cilium of Drosophila campaniform and chordotonal receptor cells. Cytoskeleton (Hoboken) 68(1):1–7. https://doi.org/10.1002/cm.20493

Liang X, Madrid J, Gartner R, Verbavatz JM, Schiklenk C, Wilsch-Brauninger M et al (2013) A NOMPC-dependent membrane-microtubule connector is a candidate for the gating spring in fly mechanoreceptors. Curr Biol 23(9):755–763. https://doi.org/10.1016/j.cub.2013.03.065

Liang X, Madrid J, Howard J (2014) The microtubule-based cytoskeleton is a component of a mechanical signaling pathway in fly campaniform receptors. Biophys J 107(12):2767–2774. https://doi.org/10.1016/j.bpj.2014.10.052

Liao M, Cao E, Julius D, Cheng Y (2013) Structure of the TRPV1 ion channel determined by electron cryo-microscopy. Nature 504(7478):107–112. https://doi.org/10.1038/nature12822

Mauthner Stephanie E, Hwang Richard Y, Lewis Amanda H, Xiao Q, Tsubouchi A, Wang Y et al (2014) Balboa binds to pickpocket in vivo and is required for mechanical nociception in Drosophila larvae. Curr Biol 24(24):2920–2925. http://dx.doi.org/10.1016/j.cub.2014.10.038

Shin JB, Adams D, Paukert M, Siba M, Sidi S, Levin M et al (2005) Xenopus TRPN1 (NOMPC) localizes to microtubule-based cilia in epithelial cells, including inner-ear hair cells. Proc Natl Acad Sci U S A 102(35):12572–12577. https://doi.org/10.1073/pnas.0502403102

Sidi S, Friedrich RW, Nicolson T (2003) NompC TRP channel required for vertebrate sensory hair cell mechanotransduction. Science 301(5629):96–99. https://doi.org/10.1126/science.1084370

Sotomayor M, Corey DP, Schulten K (2005) In search of the hair-cell gating spring elastic properties of ankyrin and cadherin repeats. Structure 13(4):669–682. https://doi.org/10.1016/j.str.2005.03.001

Suslak TJ, Watson S, Thompson KJ, Shenton FC, Bewick GS, Armstrong JD et al (2015) Piezo is essential for amiloride-sensitive stretch-activated mechanotransduction in larval Drosophila dorsal bipolar dendritic sensory neurons. PLoS One 10(7):e0130969. https://doi.org/10.1371/journal.pone.0130969

Thurm U, Erler G, Godde J, Kastrup H, Keil T, Volker W et al (1983) Cilia specialized for mechanoreception. J Submicr Cytol Path 15(1):151–155

Voelker W. Lebendbeobachtungen an kutikulaeren Reizuebertragungsstrukturen campaniformer Sensillen und Hochaufloesungs-Elektronenmikroskopie der reizaufnehmenden Sinneszellregion. PhD thesis, WestfaelischeWilhelms-Universitaet, Muenster, Germany. 1982.

Volkers L, Mechioukhi Y, Coste B (2015) Piezo channels: from structure to function. Pflugers Arch 467(1):95–99. https://doi.org/10.1007/s00424-014-1578-z

Walker RG, Willingham AT, Zuker CS (2000) A Drosophila mechanosensory transduction channel. Science 287(5461):2229–2234. https://doi.org/10.1126/science.287.5461.2229

Waterhouse AM, Procter JB, Martin DM, Clamp M, Barton GJ (2009) Jalview version 2–a multiple sequence alignment editor and analysis workbench. Bioinformatics 25(9):1189–1191. https://doi.org/10.1093/bioinformatics/btp033

Yan Z, Zhang W, He Y, Gorczyca D, Xiang Y, Cheng LE et al (2013) Drosophila NOMPC is a mechanotransduction channel subunit for gentle-touch sensation. Nature 493(7431):221–225. https://doi.org/10.1038/nature11685

Zanini D, Gopfert MC (2013) Mechanosensation: tethered ion channels. Curr Biol 23(9):R349–R351. https://doi.org/10.1016/j.cub.2013.03.045

Zhang W, Yan Z, Jan LY, Jan YN (2013) Sound response mediated by the TRP channels NOMPC, NANCHUNG, and INACTIVE in chordotonal organs of Drosophila larvae. Proc Natl Acad Sci U S A 110(33):13612–13617. https://doi.org/10.1073/pnas.1312477110

Zhang W, Cheng LE, Kittelmann M, Li J, Petkovic M, Cheng T et al (2015) Ankyrin repeats convey force to gate the NOMPC mechanotransduction channel. Cell 162(6):1391–1403. https://doi.org/10.1016/j.cell.2015.08.024

Afterword

I mentioned in the Preface that the initial motivation to write this booklet was to have an introductory material for my students. However, I realized that it is a much more rewarding process that I had a great opportunity to organize myself and also learn quite some new knowledge. As one may find out in this booklet, there have been much progresses made in the last 20 years to understand the molecular nature of mechanotransduction. However, there is still a long way to go. The question now is where we shall go from now, so here are some further questions in my mind that might be addressed in the next years:

1. *Are there other force-sensitive channels? How do they work?*

There are almost certainly other force-sensitive channels that have not been identified, for example, the channels in our auditory hair cells and those responsible for gravity sensation in fly Johnston's organ. It will be challenging to identify them because it is very likely that there are only a few of them in each cell and the mutants are just not survivable.

2. *Can we demonstrate the "gating spring" theory in vivo at molecular level?*

This needs the integration of in vivo mechanical measurement on living cells, ultrastructural analysis of the tiny mechanoreceptive organelle, and the information about the mechanical properties of all molecular components, as well as their structural organizations. I think the theory tools are necessary and probably very key!

3. *Is it possible to reconstitute a molecular or nanoscopic mechanosensor in vitro using biological materials?*

With the 3D printing and microelectronics techniques, can we build a mechanosensory device based on the knowledge learnt from the fly mechanoreceptors?

© The Author(s) 2017
X. Liang et al., *Mechanosensory Transduction in Drosophila Melanogaster*,
SpringerBriefs in Biochemistry and Molecular Biology,
DOI 10.1007/978-981-10-6526-2

4. *We should also keep in mind about the big questions to further explore the physi-ological roles of mechanotransduction in the broader sense, for example, in clinic diseases.*

Answering these four questions (and others) will be certainly exciting and challenging for the researchers in this field, including myself, in the future.